INORGANIC

REACTION MECHANISMS

THE PHYSICAL INORGANIC CHEMISTRY SERIES

Robert A. Plane and Michell J. Sienko, Editors

Physical Inorganic Chemistry	*M. J. Sienko and R. A. Plane (Cornell)*
Boron Hydrides	*W. N. Lipscomb (Harvard)*
Metal Ions in Aqueous Solution	*J. P. Hunt (Washington State)*
Inorganic Chemistry of Nitrogen	*W. L. Jolly (Berkeley)*
Inorganic Reaction Mechanisms	*J. O. Edwards (Brown)*

INORGANIC

REACTION MECHANISMS

An Introduction

JOHN O. EDWARDS

Brown University

W. A. BENJAMIN, INC.

1965　　New York　　Amsterdam

541.39
E 26

INORGANIC REACTION MECHANISMS: An Introduction

Library of Congress Catalog Card Number 64-13920
Manufactured in the United States of America

*This manuscript was received on July 31, 1963, and the
volume was published on April 10, 1964; second
printing, with corrections, Jauary 18, 1965*

*The publisher is pleased to acknowledge the assistance
of Lenore Stevens, who edited the manuscript,
Oren Hunt, who produced the illustrations,
and William Prokos, who designed the dust jacket*

W. A. BENJAMIN, INC.
New York, New York 10016

Editors' Foreword

In recent years few fields of chemistry have expanded at a rate to match that of inorganic chemistry. Aside from the stimulus afforded by the demand for new materials, a primary cause for the resurgence has been the application of physics and physical chemistry concepts to inorganic problems. As a result, both researchers active in the field and students entering the field need to become as thoroughly familiar with physical concepts as with descriptive information. However, there is presently no single point of view sufficiently general to organize the entire discipline. Instead, various points of view have arisen corresponding to the most powerful methods of attack in each research area. The synthesis of these different points of view constitutes the present series of monographs. Each monograph is contributed by an inorganic chemist active in a particular research area and reflects the methods of approach characteristic to that area. The operational procedure has been to invite able scientists to write where their interests lead them.

The series fulfills several functions. Through flexible selection of several of the monographs to supplement the introductory volume, it can be used as a textbook for an advanced inorganic chemistry course that makes full use of physical chemistry prerequisites. As a series in total, it is a reference treatise of inorganic chemistry systematized by physical principles. Finally, each monograph by itself represents a specialist's introduction to a specific research field.

It is hoped that the authors contributing to this series have succeeded in directing attention to unsolved problems and that their efforts will be repaid by continued research advances in inorganic chemistry.

M. J. Sienko
R. A. Plane

Ithaca, New York
February 1963

Preface

This book was written at the invitation of Mr. William A. Benjamin, and it is one of the series edited by Professors M. J. Sienko and R. A. Plane. As such it is designed to fit the needs of undergraduate seniors and beginning graduate students. The level is that of an introduction to the interesting and rapidly expanding field of inorganic reaction mechanisms.

In writing this book, the author assumes that the reader will have a general knowledge of inorganic and organic chemistries plus at least a year of physical chemistry, including both kinetics and thermodynamics. The first three chapters are general discussions of background material. The next six chapters are concerned with either specific types of mechanisms or specific groups of compounds. Some unsolved problems are mentioned in the last chapter.

In a book of this nature, it is possible neither to cover all of the variety of mechanisms known nor to cover any of them in great detail. A compromise in the choice of topics was made by the author, who has tried to achieve some variety plus some depth. Consequently it has been necessary to leave out many interesting studies, such as the fine work of Basolo and Wojcicki on carbonyl exchanges.

This book could not have been completed without the help of many people. Professors M. J. Sienko and R. A. Plane, of Cornell, Professor Cooper Langford, of Amherst, Mr. Joseph Epstein, of the

Army Chemical Center, and Professor Khairat M. Ibne-Rasa, of Brown, have made suggestions as to the scientific content. Miss Lois Nyberg, Mrs. Ruth C. Edwards, and particularly Mrs. Alda O. Kirsis have helped with the wording and grammar. Brown University has made available the necessary time, and Professors John Ross and Richard L. Carlin have significantly helped by taking over many of my lectures. For their forbearance while his mind was solely on the book, the author is grateful to his family and his students.

JOHN O. EDWARDS

Providence, Rhode Island
May 1963

Contents

INORGANIC

REACTION MECHANISMS

I

Introduction

1-1 THE SUBJECT OF THIS BOOK

The field of inorganic reaction mechanisms is not a new one, for as early as the turn of the century several chemists were studying the kinetics of oxidation-reduction and acid-base reactions. In the succeeding years, particularly around 1930, acid-base catalysis was investigated rather carefully. These early investigations were carried out mostly by physical chemists, who employed only kinetics on simple aqueous mixtures and were interested mainly in testing the theories current at that time. During the same period, the organic chemists were becoming interested in detailing the pathways that atoms take during the course of a reaction. The now popular field of physical organic chemistry thus sprang into being, and interest in organic mechanisms is currently widening into the area of biochemical reactions.

In recent years, the inorganic chemists have become aware of the large array of systems available to them for study, both by the kinetic methods of the physical chemist and by the detailed stereochemical techniques of the organic chemist. It would be unfair, however, to assume that the relatively late entry of the pure inorganic chemist into the area of mechanisms showed a lack of interest. In fact, until the complex pattern of structures and bondings in inorganic molecules was partially resolved, there was little that the inorganic chemist could do in the area of mechanisms except for reaction kinetics. The nature of bonding in inorganic compounds is not yet fully resolved, and there are many structures to be worked out. There is no doubt, however, that the field of reaction mechanisms for inorganic compounds is receiving considerable attention at present.

Before discussion of actual cases, there are several points that must be mentioned to establish between the reader and the author a common set of definitions and symbols. This chapter deals with such points.

The *inorganic reactions* with which we shall deal involve compounds other than those containing carbon-hydrogen or carbon-carbon bonds at the reaction site. Because of the quality and quantity of previous work on organic compounds, it will be necessary to use organic examples for comparative purposes. Also, it has often been necessary (for stability, structural, electronic, or analytical reasons) for the inorganic chemist to use organic substituents in his systems, and such cases will be included here. Primarily, however, the cases will be noncarbon compounds.

By *reaction mechanism*, we mean the detailed, stepwise pattern of atomic and electronic motions that take place while reactants change to products. Heavy reliance on chemical kinetics will be necessary, because in inorganic systems many rapid equilibria are possible before the transition state. Other clues to mechanism are obtained through isotope-tracer studies, stereochemical investigations, medium effects (including both solvents and electrolytes), and linear free-energy relations.

1-2 STOICHIOMETRY

In all studies of mechanism, one of the first and certainly most important pieces of data needed is the *stoichiometry* of the reaction. There are two ways to interpret this word and both are correct. The more limited way, and this is the one we shall employ here, is that stoichiometry refers to the number of particles (atoms, molecules, and ions) in the balanced chemical equation representing the chemical reaction. In many cases, where more than a single reaction occurs, stoichiometry may refer more broadly to how much reactant(s) goes to what product(s) and to the extent to which the various reactions compete.

The need for correct stoichiometry (equation balancing) is a point that most students do not fully recognize. One cannot quantitatively treat or discuss a reaction without first having balanced the equations for this reaction. It is also worth noting that identification of all products of the reaction can be useful in elucidation of mechanism. A single product suggests a single transition state involving atom transfer, whereas multiple products usually occur where the mechanism involves molecular fragmentation and competing transition states.

1-3 EQUILIBRIA

The relationship between the yield in a reaction at infinite time and the yield at some specific time during the course of the reaction is one that has not been carefully handled in some cases. Since this interrelation between thermodynamics and kinetics will be used in the chapters following, it is appropriate that the matter be covered here.

Thermodynamics is the area of physical chemistry that deals with the spontaneity of a chemical reaction. The first step in considering the potential yield of a reaction is to set up an expression of the type

$$K_c = \frac{[M]^m[N]^n\cdots}{[A]^a[B]^b\cdots} \tag{1-1}$$

where K_c is a temperature-dependent constant, brackets denote concentrations, and the other symbols are derived from the balanced chemical equation

$$a\text{A} + b\text{B}\cdots \rightleftharpoons m\text{M} + n\text{N}\cdots$$

Only at equilibrium will Eq. (1-1), called the *equilibrium-constant expression*, apply, for only in that state does the given ratio of concentrations become equal to the constant K_c. This constant can be calculated from the free energies of the reactants and products.

The equilibrium-constant formulation, which was first recognized by Guldberg and Waage in 1864, is commonly called the *law of mass action*. These workers stated that the extent of a reaction is proportional to the molar concentration of each reacting substance raised to a power equal to the number of molecules of that substance taking part in the reaction. A more general statement is: the product of the concentrations of the products of a reaction, each raised to a power equal to the number of partaking molecules, divided by the product of the concentrations of the reactants, each raised to its analogous power, is equal to a constant at any given temperature.

For example, the reaction

$$H_2 + I_2 \rightleftharpoons 2HI$$

in the gas phase reaches an equilibrium position corresponding to the mass expression

$$K_c = \frac{[HI]^2}{[I_2][H_2]} \tag{1-2}$$

where K_c has the value 45.9 at 490°C. Using this value, one can calculate percentage yields and the like at this temperature. It is important to remember that most equilibrium constants have dimensions; thus the value obtained for K_c will usually depend on the concentration units employed.

The equilibrium-constant expression has been given in terms of concentrations. Since the *activity* of a material (which might be defined as its effective concentration) is often different from its concentration, it is more strictly correct to use activities in the equilibrium expression. Use of concentrations may be satisfactory in certain cases, as for neutral molecules in dilute phases; in other cases, such as with ionic solutes, it is imperative that activities be employed.

Since the concentration of water is in great excess in most aqueous solutions studied, the activity of water is constant. This activity can, therefore, be placed in the constant of mass expression; one does not find the water concentration (or activity) expressed as such along with other concentration terms. An example will be given below.

Consider also the solubility-product expression for lead sulfide

$$K_{sp} = [Pb^{++}][S^=] \tag{1-3}$$

The reaction is

$$PbS(s) \rightleftharpoons Pb^{++} + S^=$$

where the solid material is in equilibrium with the aqueous ions.

Since the activity of a solid material is constant and independent of the amount present, the expression

$$K' = \frac{[\text{Pb}^{++}][\text{S}^=]}{A_{\text{PbS}}} \tag{1-4}$$

where A_{PbS} is the lead sulfide activity, reduces to the K_{sp} expression [Eq. (1-3)] by multiplication of the two constants. When one remembers that the concentration (mass per unit volume) and therefore the activity is related to the density of a solid and that this remains constant and independent of the surrounding solution at any temperature, one can see why it is unnecessary to use any more complicated expression than the K_{sp} relation for treating solubility equilibria. In general, whenever the activity of one constituent in a reaction is unchanged over a range of conditions, it may be included in the constant K of the equilibrium expression.

1-4 REACTION RATE

In the study of a reaction, we must consider both the equilibrium (final) state and the velocity of approach to this state. We talk about reactions in terms of three constants: the equilibrium constant K_c, the rate constant in the forward direction k_f, and the reverse-rate constant k_r. For a reaction far from equilibrium (i.e., negligible reverse reaction) the velocity is given by a *rate law* of the type

$$v = k_f([\text{X}]^p[\text{Y}]^q \cdots) \tag{1-5}$$

where X and Y are various chemical species, p and q are exponents obtained from experimental data, v is the rate of reaction, and k_f is the temperature-dependent rate constant characteristic of this reaction. It is not to be assumed that X and Y always represent the reactants in the stoichiometric equation, and it is indeed rare when p and q are the stoichiometric powers for related species in the equilibrium-constant expression. At the present stage of chemical kinetics, one must rely on experimentation to obtain the correct rate law for a reaction.

At equilibrium, the transfer of atoms in the forward and reverse directions still takes place even though no net concentration change is observed. There must, therefore, be a relation between the two rate constants and the equilibrium constant. This relation

$$K_c = \frac{k_f}{k_r} \tag{1-6}$$

is known as the *law of microscopic reversibility*. We shall assume it to be universally applicable although it has been demonstrated experimentally only a few times. One of the reactions for which Eq. (1-6) has been shown to hold is

$$H_3AsO_4 + 3I^- + 2H^+ \rightleftharpoons H_3AsO_3 + H_2O + I_3^-$$

The rate law for formation of I_3^- is

$$\frac{d[I_3^-]}{dt} = k_f[H_3AsO_4][I^-][H^+] \tag{1-7}$$

Using the definition of K_c [Eq. (1-1)] and the equation for microscopic reversibility [Eq. (1-6)], we find that

$$K_c = \frac{[H_3AsO_3][I_3^-]}{[H_3AsO_4][I^-]^3[H^+]^2} = \frac{k_f}{k_r} \tag{1-8}$$

or

$$\frac{k_r[H_3AsO_3][I_3^-]}{[I^-]^2[H^+]} = k_f[H_3AsO_4][I^-][H^+] \tag{1-9}$$

At equilibrium, the rate of formation of I_3^- (v_f) must equal the rate of removal of I_3^- (v_r); this may be written

$$v_f = v_r \tag{1-10}$$

If there exists no other mechanism for the process, then these rate laws are valid at all times (not only near but also far from equilibrium). We have seen in Eq. (1-7) that

$$v_f = \frac{d[I_3^-]}{dt} = k_f[H_3AsO_4][I^-][H^+]$$

Thus it follows that Eq. (1-11)

$$v_r = \frac{-d[I_3^-]}{dt} = \frac{k_r[H_3AsO_3][I_3^-]}{[I^-]^2[H^+]} \tag{1-11}$$

is the correct rate law for the reverse reaction. One should note, as discussed in Section 1-3, the absence of water-concentration terms both in the equilibrium-constant expression and in the rate laws.

The experimental results obtained in the kinetic study of this reaction showed that the forward- and reverse-rate laws were congruent (i.e., in harmony with each other) even though measured under widely different conditions. The value of K_c was known from studies of equilibria to be 6.3, and the ratio between the two experimental rate constants proved to be 6.7. For the field of kinetics, this is remarkable agreement.

The correctness of the assumption that a reaction is proceeding even while at equilibrium has been demonstrated for the same system. By use of radioactive tracers, it was shown that individual arsenic atoms change oxidation states with the same rate laws and rate constants even though no net concentration changes occur.[23] These results are, of course, a beautiful demonstration of the dynamic nature of chemical equilibrium and of the validity of the law of microscopic reversibility.

The value of the exponent on a concentration term in a rate law is spoken of as the *order* for this component. For example, in the rate law

$$\frac{-d[I_3^-]}{dt} = \frac{k_r[H_3AsO_3][I_3^-]}{[I^-]^2[H^+]}$$

[Eq. (1-11)], the reaction is first-order each in arsenious acid and triiodide ion concentration, it is inverse second-order in iodide ion concentration, and it is inverse first-order in hydrogen ion concentration.

For a simple rate law with no concentration terms in the denominator, the over-all order of the reaction is the sum of the exponents. For the rate law

$$\frac{d[A]}{dt} = k[B]^2[C] \tag{1-12}$$

the reaction is spoken of as third-order. When there are concentration terms in the denominator of a rate law, it is probably unwise to speak of an over-all order, since the value obtained by summation of exponents is in no way indicative of the true complexity of the rate law.

One often hears the term *bimolecular reaction* when the over-all order is two. Although such an expression may be correct in certain cases, *molecularity* has a different meaning from order. The number of particles (molecules, ions, atoms) interacting in any given step of a reaction is the molecularity of this step. Thus, we have unimolecular, bimolecular, and occasionally termolecular steps being postulated in reaction mechanisms.

A third-order reaction does not necessarily mean that a triple collision takes place in the reaction. The difference between molecularity and order can be illustrated with the following example. In a three-car collision, two cars usually collide first, and later a third car collides with the first two. Similar stages occur in the reactions of molecules, as is shown in the following steps that lead to the activated complex for the arsenic acid and iodide ion reaction.

$$
\begin{array}{c}
O \\
| \\
HO{-}As{-}OH + H^+ \rightleftharpoons HO{-}As^+ + H_2O \\
| \\
O \\
H
\end{array}
\qquad
\begin{array}{c}
O \\
| \\
\\
| \\
O \\
H
\end{array}
$$

$$
\begin{array}{c}
O \\
| \\
HO{-}As^+ + I^- \rightleftharpoons \\
| \\
O \\
H
\end{array}
\;
\left[
\begin{array}{c}
O \\
| \\
HO{-}As{:}I \\
| \\
O \\
H
\end{array}
\right]^0
\;
\rightarrow products
$$

Here, a third-order reaction is made up of two bimolecular steps. It is doubtful that any step in a reaction mechanism has a molecularity greater than three.

It is also possible to relate the difference between order and molecularity to their respective ultimate sources. An order is derived from kinetic experiments, whereas any molecularity is a theoretical construct (since postulated mechanisms are theories).

1-5 TRANSITION-STATE THEORY

Because of its usefulness, the terminology of the transition-state theory[8] of chemical kinetics will be used in this book. This theory is based on a model that assumes that the reactants are in equilibrium with a "hot" chemical species called the *activated complex* and that the activated complexes for all reactions break down to products with the same universal rate constant. The amount of energy necessary to form the activated complex is the *activation energy*, and the point of highest energy on the path from reactants to products is called the *transition state*. The phrases "activated complex" and "transition state" are often used interchangeably; however, there is a difference. The transition state is an energy state, and the activated complex is the chemical species present in this energy state. The symbol $\neq$ will be used here to represent the activated complex.

The formulation of thermodynamics has been applied by Eyring[8] to the hypothesized equilibrium between reactants and activated complex. The hypothetical equilibrium constant is symbolized $K^{\neq}$. The quantities $\Delta F^{\neq}$, $\Delta H^{\neq}$, and $\Delta S^{\neq}$ are called the activation thermodynamic parameters (free energy of activation, enthalpy of activation, and entropy of activation, respectively). They are defined by the equations

$$\Delta F^{\neq} = -RT \ln K^{\neq} \tag{1-13}$$

$$\Delta H^{\neq} = RT^2 \frac{d(\ln K^{\neq})}{dT} \tag{1-14}$$

$$\Delta S^{\neq} = \frac{\Delta H^{\neq} - \Delta F^{\neq}}{T} \tag{1-15}$$

where T is the absolute temperature and R is the gas constant. The relationship between rate constant and free energy of activation is expressed by Eq. (1-16)

$$k_f = \frac{kT}{h} \exp\left(-\frac{\Delta F^{\neq}}{RT}\right) \tag{1-16}$$

where k is the Boltzmann constant, h is Planck's constant, and the other symbols have their previously given meanings.

As reactants pass over to products, the internal energy of the particles first increases and then decreases. That path on an energy surface that requires the least energy of all possible paths from reactants to products is called the *reaction coordinate*. It is convenient to plot energy against reaction coordinate; this type of plot, an example of which is given in Figure 1-1, is called a *reaction-coordinate diagram*. Since we rarely know much about the internal energy of a reacting system, the plots often show free energy against reaction coordinate.

In this diagram, state A represents the energy of the reactants, B is a barrier between A and C, C is the energy minimum resulting from an intermediate before the activated complex, D is the transition state, E is the energy minimum resulting from an intermediate after the transition state, F is a small energy barrier between E and G, and G is the energy of the final products. The difference in energy between states A and D is the activation energy

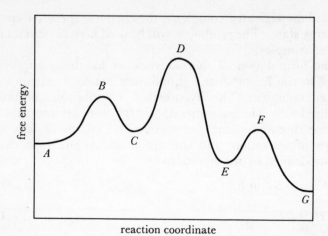

Figure 1-1 *A reaction-coordinate diagram. The free energy of a reacting system is plotted against the hypothetical reaction coordinate, which is the path the system takes in going from reactants to products.*

in the forward direction; in the reverse direction the activation energy is the difference between G and D. An *intermediate* is a particle whose energy is a minimum in the curve on the reaction-coordinate diagram and as such is potentially isolable. An activated complex such as D (or the subsidiary ones, B and F) has a lifetime of only a few molecular collisions and is not isolable.

1-6 ORDER AND REACTION MECHANISM

Certain consequences of the rate law are useful in the consideration of reaction mechanism. They stem from the kinetic orders obtained in the experimental determination of rate as a function of concentration. In brief, the principal consequence is this: the species that appear in the rate law are those present in the activated complex. Irrespective of postulated mechanism, the activated complex for the arsenate oxidation of iodide must contain one arsenic atom plus one iodine atom, and the complex must be electrically neutral. This follows from both the forward-rate law [Eq. (1-7)]

$$v_f = k_f[H_3AsO_4][I^-][H^+]$$

and the reverse-rate law [Eq. (1-11)]

$$v_r = k_r \frac{[H_3AsO_3][I_3^-]}{[I^-]^2[H^+]}$$

It will be noted as a second consequence that when one or more species appears in the denominator, as in the reverse-rate law above, these atoms and charges are subtracted from those present in the numerator; chemically speaking, this means that there are rapid equilibria before the rate-determining step.

Another consequence of the rate-law formulation is that nothing is known of the number of solvent molecules present in the activated complex; this stems from our previous conclusion that the activity of solvent is constant and therefore is lumped into the rate constant.

Thus we can draw several valuable conclusions about this reaction: the activated complex has the composition $AsIO_2 \cdot xH_2O$, and there are rapid equilibria before the activated complex when approached from the reverse direction. The absence of inverse orders in a rate law does not, of course, rule out the presence of rapid equilibria.

1-7 SYMBOLS

In Table 1-1, the symbols to be used in this book are listed along with their meanings.

Supplementary Reading

Students who wish to go further into the subject of inorganic mechanisms will, in view of the limited coverage of this book, need to go to the literature. Listed in the references below are sources of more detailed reading. The books on chemical kinetics[1-10] usually have some space devoted to mechanisms; Chapter 11 in Ref. 3 is particularly noteworthy. The large amount of work in the area of organic reaction mechanisms is reflected in the number of books dealing with this area.[14-20] For reading in inorganic mechanisms, Refs. 12 and 13 are excellent, as are the reviews of Ref. 22. In the main, however, the student will have to rely on articles in periodicals for a deeper insight into inorganic mechanisms. The article by Taube[21] in J. Chem. Educ. is especially recommended to all students and teachers who are interested in inorganic kinetics and mechanisms.

Table 1-1   *Symbols and their meanings*

k	Rate constant
k_f	Rate constant in the forward direction
k_r	Rate constant in the reverse direction
v	Rate (velocity) of reaction
[A]	Concentration of species A
k	Boltzmann's constant
h	Planck's constant
R	Gas constant
K_c	Concentration equilibrium constant
K	Thermodynamic equilibrium constant
ΔF	Free-energy change
ΔH	Enthalpy change
ΔE	Energy change
ΔS	Entropy change
$K^{\neq}$	Equilibrium constant (reactants to activated complex)
$\Delta F^{\neq}$	Activation free energy
E_a	Energy of activation
$\Delta H^{\neq}$	Enthalpy of activation
$\Delta S^{\neq}$	Entropy of activation
$\neq$	Activated complex
T	Absolute temperature
t	Time

References

Kinetics

1. L. S. Kassel, *Kinetics of Homogeneous Gas Reactions*, ACS Monograph No. 57, Chemical Catalog Co., New York, 1932.
2. E. A. Moelwyn-Hughes, *The Kinetics of Reactions in Solutions*, Clarendon Press, Oxford, 2d ed., 1947.
3. A. A. Frost and R. G. Pearson, *Kinetics and Mechanism*, 2d ed., Wiley, New York, 1961.
4. C. N. Hinshelwood, *The Kinetics of Chemical Change*, Clarendon Press, Oxford, 1947.
5. N. N. Semenov, *Some Problems in Chemical Kinetics and Reactivity*, Princeton University Press, Princeton, N.J., 1958, Vols. 1, 2.
6. E. S. Amis, *Kinetics of Chemical Change in Solution*, Macmillan, New York, 1949.
7. K. J. Laidler, *Chemical Kinetics*, McGraw-Hill, New York, 1950.
8. S. Glasstone, K. J. Laidler, and H. Eyring, *The Theory of Rate Processes*, McGraw-Hill, New York, 1941.
9. S. W. Benson, *The Foundations of Chemical Kinetics*, McGraw-Hill, New York, 1960.

10. *Investigation of Rates and Mechanisms of Reactions*, in S. L. Friess, E. S. Lewis, and A. Weissberger (eds.), *Technique of Organic Chemistry*, 2d ed. (revised), Wiley-Interscience, New York, 1961, Vol. VIII (2 parts).
11. Symposium on Anomalies in Reaction Kinetics, *J. Phys. Colloid Chem.*, **55**, 763–1104 (1951). A group of interesting papers and discussions collected in one issue of the journal.

Mechanism
12. F. Basolo and R. G. Pearson, *Mechanisms of Inorganic Reactions*, Wiley, New York, 1958. This book deals primarily with coordination compounds.
13. R. P. Bell, *The Proton in Chemistry*, Cornell University Press, Ithaca, N.Y., 1959. A book covering equilibria, kinetics, and mechanisms of protonic reactions.
14. E. R. Alexander, *Principles of Ionic Organic Reactions*, Wiley, New York, 1950.
15. J. Hine, *Physical Organic Chemistry*, McGraw-Hill, New York, 2d ed., 1962.
16. M. S. Newman, *Steric Effects in Organic Chemistry*, Wiley, New York, 1956.
17. C. K. Ingold, *Structure and Mechanisms in Organic Chemistry*, Cornell University Press, Ithaca, N.Y., 1953.
18. L. P. Hammett, *Physical Organic Chemistry*, McGraw-Hill, New York, 1940.
19. G. E. K. Branch and M. Calvin, *The Theory of Organic Chemistry*, Prentice-Hall, Englewood Cliffs, N.J., 1941.
20. E. S. Gould, *Mechanisms and Structure in Organic Chemistry*, Holt, Rinehart, and Winston, New York, 1960.

Articles
21. H. Taube, *J. Chem. Educ.*, **36**, 451 (1959).
22. Review articles on recent results (references, summaries, and discussions) are to be found in *Ann. Rev. Phys. Chem.* These cover both kinetics and mechanism.
23. J. N. Wilson and R. G. Dickinson, *J. Am. Chem. Soc.*, **59**, 1358 (1937).

2

Binding Strength and Reaction Rates: Acids, Bases, and Solvents

The rate law for a chemical reaction is important to the elucidation of its mechanism in that it defines certain gross features such as the constitution of the activated complex. However, to learn further details about the mechanism, it is necessary to examine such other types of information as electronic and steric influences on rate and the influence of acid and base on rate. Linear free-energy relations[1-3] are a guide to the understanding of such electronic effects and are tied in with the quantitative treatment of acid and base theories. The problem of acid and base catalysis is also important to the study of mechanisms in its own right. For these reasons, this chapter will deal with acid-base theories[4] and the next chapter with linear free-energy relations.

2-1 BRÖNSTED THEORY

All acid-base theories are classification schemes† that tell what an acid is and what a base is. The theory of J. N. Brönsted[5] defines

---

† In the sense that many scientists use the word theory, the Brönsted and Lewis contributions are not theories; they really are definitions. For example, Brönsted *defines* an acid as a proton donor.

14

an acid as a proton donor [Eq. (2-1)] and a base as a proton acceptor [Eq. (2-2)].

$$HA \rightarrow H^+ + A^- \tag{2-1}$$

$$H^+ + B \rightarrow HB^+ \tag{2-2}$$

As indicated by Eq. (2-1), the acid HA gives up a proton; in Eq. (2-2), the base B accepts a proton. Because of its exceptionally large ratio of charge to radius, the proton does not exist as such in a condensed phase; it has been estimated that the hydration energy of the proton is more than a quarter million calories per mole. Thus, the equation for a real acid-base reaction is

$$HA + B \rightleftharpoons HB^+ + A^- \tag{2-3}$$

in which the acid donates a proton to the base to form a new acid HB^+ and a new base A^-. The particles HA and HB^+ are called the conjugate acids of bases A^- and B, respectively; particles B and A^- are called the conjugate bases of acids HB^+ and HA, respectively. Amphoterism is the ability of a single particle either to donate or to accept a proton depending on the circumstances.

The strength of a Brönsted acid is measured by the acid's ability to donate a proton to the solvent molecule. Thus, in aqueous solution, the ionization constant K_a is defined by the mass expression

$$K_a = \frac{[H^+][A^-]}{[HA]} \tag{2-4}$$

where H^+ is the solvated proton, HA is the acid, and A^- is its conjugate base. The value of K_a is the quantitative measure of acid strength. For the measure of base strength, the analogous expression

$$K_b = \frac{[BH^+][OH^-]}{[B]} \tag{2-5}$$

is employed in aqueous solution.

2-2   SPECIFIC-ACID CATALYSIS

Reactions with rate laws of the type

$$v = k[S][H^+]^n \tag{2-6}$$

where S is a substrate and n is the order in hydrogen ion concentration, are often observed. According to Bell,[6] a catalyst may be defined as any particle that appears in the rate law with an order larger than its coefficient in the stoichiometric equation. Thus, if the value of n is larger than the number of hydrogen ions in the equation for the reaction, the reaction is said to be acid-catalyzed.

Specific-acid catalysis is enhancement of a reaction rate by the solvated proton only. The rate should depend solely on the hydrogen ion concentration and not at all on the amount of undissociated acid present in the solution. For example, if the rate of a reaction is enhanced by hydronium ion in water but not by acetic acid molecules, the reaction is said to be specific-acid-catalyzed. In other words, the rate is a function of pH but not of the amount or chemical nature of undissociated acid. In Table 2-1, some data illustrating this type of catalysis are presented; the hydrogen ion concentration in these solutions was measured by the conductivity method, and the reactions being followed were (I) the hydrolysis of methyl acetate and (II) the inversion of cane sugar. There is a close parallel between the hydrogen ion concentration and the rates of the two reactions.

Table 2-1   *Relative conductivities and catalytic effects of acids[a]*

Acid	Conductivity	Reaction I[b]	Reaction II[c]
HCl	(100)	(100)	(100)
HBr	101	98	111
HNO_3	99.6	92	100
H_2SO_4	65.1	73.9	73.2
Cl_3CCO_2H	62.3	68.2	75.4
Cl_2CHCO_2H	25.3	23.0	27.1
Oxalic	19.7	17.6	18.6
$ClCH_2CO_2H$	4.90	4.30	4.84
Formic	1.67	1.30	1.53
Lactic	1.04	0.90	1.07
Acetic	0.424	0.345	0.400
Isobutyric	0.311	0.286	0.335

[a] All data are from aqueous 1-N solutions of the acids (Ostwald, 1884[7]). Values for hydrochloric acid are arbitrarily set at 100 in each case; other values are relative to them.

[b] Rate of hydrolysis of methyl acetate.

[c] Rate of inversion of cane sugar.

The mechanism of a specific-acid-catalyzed reaction involves a rapid equilibration of the substrate with hydrogen ion to form a protonated species, which then in a slow, rate-determining step reacts to give products. In the case of methyl acetate hydrolysis, the mechanism can be briefly written

$$CH_3C\underset{OCH_3}{\overset{O}{\diagup}} + H_3O^+ \rightleftharpoons CH_3C\underset{OCH_3}{\overset{OH^+}{\diagup}} + H_2O$$

$$CH_3C\underset{OCH_3}{\overset{OH^+}{\diagup}} + H_2O \rightarrow CH_3C\underset{OH}{\overset{OH^+}{\diagup}} + HOCH_3$$

A reaction is said to be specific-base-catalyzed when it is catalyzed by the hydroxide ion in aqueous solution but not by basic anions such as borate ion nor by undissociated bases such as ammonia. The mechanism of such catalysis involves a rapid-equilibrium loss of a proton by substrate followed by a slow, rate-determining step that forms the product. This type of mechanism may be represented by the steps

$$HS + OH^- \rightleftharpoons S^- + H_2O$$

$$S^- \rightarrow products$$

2-3   GENERAL-ACID CATALYSIS

The other common type of proton catalysis is observed when hydrogen ions are transferred from a Brönsted acid to a substrate in the transition state.[2] In such cases, any acidic molecule or ion can act to transfer the proton to the substrate; therefore a multiterm rate law will be observed. Experimentally it is found that the rate increases with hydrogen ion concentration as in specific-acid catalysis, but the rate also increases with the amount of buffer acid at constant hydrogen ion concentration. Mechanistically, this can be formulated as

$$HA + S \rightarrow SH^+ + A^-$$

$$SH^+ \rightarrow products$$

where the first step is rate-determining and the second is rapid, and where any acidic species present in solution can act as HA.

The rate law observed in aqueous solution for this type of acid catalysis is

$$v = k_1[S][H_3O^+] + k_2[S][HA] + k_3[S] \tag{2-7}$$

where the first term represents catalysis by hydronium ion, the second term represents catalysis by molecular acid (such as undissociated acetic acid), and the third term represents catalysis by solvent molecules. A general formulation for such a rate law is

$$v = [\text{S}] \left(\sum k_{\text{HA}}[\text{HA}] \right) \tag{2-8}$$

with HA here representing all possible proton donors in solution. General-acid catalysis is also observed in other solvent systems.

General-base catalysis is observed when all proton acceptors will catalyze a reaction by removing a proton from the substrate in the transition state. The generalized rate law

$$v = [\text{SH}] \left(\sum k_{\text{B}}[\text{B}] \right) \tag{2-9}$$

is obeyed in such cases.

2-4 BRÖNSTED EQUATION

The experiments that proved crucial to the Brönsted theory of acids and bases (as compared to the earlier Arrhenius theory) were those that showed the relation of rates of general-acid catalysis [k_{HA} of Eq. (2-8)] to the abilities of proton acids to form hydronium ion in solution [K_a of Eq. (2-4)]. It was known that the values of k_{HA} correlated with the ionization constants K_{HA}. Brönsted[2] demonstrated that this relation has the form

$$k_{\text{HA}} = G_{\text{A}} K_{\text{HA}}^{\alpha} \tag{2-10}$$

where G_{A} and α are constants characteristic of the reaction under study. In similar fashion, the relation between the rate constants k_{B} for the various terms of a rate law for general-base catalysis [Eq. (2-9)] and the appropriate basicity constants K_{B} is

$$k_{\text{B}} = G_{\text{B}} K_{\text{B}}^{\beta} \tag{2-11}$$

where G_{B} and β are again constants for the particular reaction.

Both in general-acid catalysis and in general-base catalysis, the proton is only partially transferred in the activated complex. If transfer were complete, then the values for the exponents α and β would be unity, since ionization (as measured by K_{HA} and K_{B}) involves complete dissociation. If transfer were negligible, then the values of α and β would be 0; there would be no observable acid-base catalysis in this case. In actual practice, the values of α and β lie between 0 and 1, and the magnitude of the value may be taken as a measure of the degree of proton transfer in the activated complex.†

Some results that exemplify the Brönsted treatment are found in the decomposition of nitramide,[8]

$$H_2NNO_2 \rightarrow H_2O + N_2O$$

which is subject to general-base catalysis. Numerical data are given in Table 2-2 and some of these data (Nos. 2–19) are recast into a Brönsted plot in Figure 2-1. The logarithmic correlation

Table 2-2 *Rate constants for general-base catalysis of nitramide decomposition*[a]

Base	No.[b]	k_B, liter mole^{-1} min^{-1}	K_B
Monohydrogenphosphate ion	1	86	2.0×10^{-7}
Succinate ion	2	1.8	4.8×10^{-9}
Pyridine	3	4.6	2.3×10^{-9}
Dimethylaniline	4	2.7	1.6×10^{-9}
Malate ion	5	0.72	1.5×10^{-9}
Quinoline	6	1.9	8.3×10^{-10}
Propionate ion	7	0.65	7.5×10^{-10}
Acetate ion	8	0.50	5.5×10^{-10}
Aniline	9	0.531	5.0×10^{-10}
Tartrate ion	10	0.165	2.7×10^{-10}
Oxalate ion	11	0.104	2.2×10^{-10}
Phenylacetate ion	12	0.23	1.9×10^{-10}
Benzoate ion	13	0.19	1.5×10^{-10}
Hydrogensuccinate ion	14	0.320	1.5×10^{-10}
Formate ion	15	8.2×10^{-2}	4.8×10^{-11}
Hydrogenmalate ion	16	7.65×10^{-2}	2.5×10^{-11}
Hydrogentartrate ion	17	3.63×10^{-2}	1.03×10^{-11}
Salicylate ion	18	2.1×10^{-2}	1.0×10^{-11}
Hydrogenphthalate ion	19	2.9×10^{-2}	8.3×10^{-12}
Dihydrogenphosphate ion	20	7.9×10^{-3}	1.1×10^{-12}
Dichloracetate ion	21	7×10^{-4}	2.0×10^{-13}
Water	22	$\sim 8 \times 10^{-6}$	$\sim 2 \times 10^{-16}$

[a] Data from Ref. 8; no corrections have been made for statistical factors.
[b] Correspond to numbered points on Figure 2-1.

† An interesting discussion of the interrelation of hydrogen bonding and general-acid catalysis has been presented by J. E. Gordon [*J. Org. Chem.*, **26**, 738 (1961)].

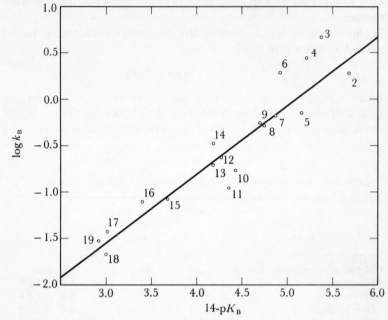

Figure 2-1   *A Brönsted plot for the general-base-catalyzed decomposition of nitramide.*

between rate constants and ionization constants for various bases is apparent.

2-5   STATISTICAL CORRECTIONS

Both rate constants and equilibrium constants depend on the molecular symmetries of the particles involved. For example, the symmetry number for proton transfer by the acid NH_4^+ is twelve, since there are four equivalent protons and with any one proton chosen there are three possible equivalent positions (obtained by rotation) for the other three protons. By way of contrast, the symmetry numbers for proton transfer involving RNH_3^+, $R_2NH_2^+$, and R_3NH^+ are three, two, and three, respectively. It can be seen from these examples that the symmetry number is equal to the number of equivalent configurations obtained by molecular rotations only.

For symmetry corrections in a particular reaction, the case of the ionization constants of a long-chain dicarboxylic acid is a good example. Assuming no interaction between the two ends, one would predict the first ionization constant K_1 to be four times as large as the second constant K_2. This can be confirmed by comparing K_1 and K_2 with the ionization constant K_0 for a long-chain mono-carboxylic acid of identical chemical nature (same intrinsic acidity of the proton and same intrinsic basicity of the carboxylate anion). Since the dicarboxylic acid has two equivalent protons and its conjugate base has only one basic site, it follows that K_1 must be twice as large as K_0. Using similar reasoning, it can be seen that $K_2 = \frac{1}{2}K_0$. Therefore $K_1 = 4K_2$ on the basis of symmetry alone. The need for statistical correction in general-acid and general-base catalysis was pointed out by Brönsted[2,8]; a portion of the scatter seen in Figure 2-1 results from the lack of application of statistical corrections. Recently, Benson[9] has discussed in some detail symmetry corrections for constants.

In Table 2-3, constants for reactions involving chelate complexes of hydroxyanions with glycerol $(R = H)$ and its monomethyl ether $(R = CH_3)$ are given. All data are for reactions of the type

$$^{\ominus}M\!\!\begin{array}{c}\diagup OH \\ \diagdown OH\end{array} + \begin{array}{c}HO\diagdown CH_2 \\ | \\ HO\diagup CH \\ | \\ CH_2OR\end{array} \underset{k_r}{\overset{k_f}{\rightleftharpoons}} {}^{\ominus}M\!\!\begin{array}{c}\diagup O\diagdown CH_2 \\ | \\ \diagdown O\diagup CH \\ | \\ CH_2OR\end{array} + 2H_2O$$

Table 2-3   *Statistical factors for glycol complexes[a]*

Hydroxyanion	Constant	$R = H$[b]	$R = CH_3$[b]	Ratio[c]
Borate	formation	16.0	7.5	2.13
Phenylboronate	formation	19.7	8.45	2.33
Arsenite	formation	1.15	0.66	1.74
Tellurate	formation	79.6	30.2	2.63
Tellurate	forward rate	4.99×10^{-8}	2.06×10^{-8}	2.42
Tellurate	reverse rate	0.63×10^{-9}	0.68×10^{-9}	0.92

[a] For discussion of values, see text.
[b] In the compound $HOCH_2CHOHCH_2OR$.
[c] Constant in third column divided by constant in fourth column.

where M denotes the rest of the hydroxyanion and K_c is the equilibrium constant. When R = H, it can be seen that there are two different ways for a complex to form, for there are two identical ends. Thus, K_c and k_f should be twice as large as the values obtained when R = CH_3, assuming, of course, that all other factors are negligibly different, as seems reasonable. The observed ratios are in agreement. Also in good accord is the ratio for k_r values, which is predicted to be 1 since the complex can only break down in one way.

In general, statistical corrections to rate and equilibrium constants are small compared to the effects of electronic and steric factors; the corrections should, however, be applied whenever possible.

2-6 LEWIS THEORY

One other theory of acids and bases is widely employed in the discussion of mechanism. This theory, first put forward by Lewis[10] in 1923, classifies acids and bases according to their electronic behaviors. An acid is defined as an electron-pair acceptor, a base as an electron-pair donor; the neutralization process is characterized by the formation of a covalent bond. For example, boron trifluoride is a Lewis acid (because of the unfilled p orbital on boron), trimethylamine is a base (there is a nonbonded pair of electrons on the nitrogen), and neutralization is the formation of the adduct F_3B—$N(CH_3)_3$, which has a new sigma bond between boron and nitrogen. Some other Lewis acids are $AlCl_3$, SO_3, and Ag^+; some Lewis bases are Cl^-, PH_3, C_6H_6, and olefins. In general, particles that are Lewis bases may also be Brönsted bases since electron-pair donation to the proton is the actual mechanism for basicity in the Brönsted sense. For acids, the two theories are notably divergent.

The observation that aluminum trichloride and mercuric ion can catalyze the same types of reaction as are subject to proton-acid catalysis was one of the foundations of the Lewis theory. This is clearly exemplified by the work of Bell and Skinner[11] on the depolymerization of paraldehyde.

$$\rightleftarrows 3CH_3CHO$$

The reaction proceeds slowly in the absence of a catalyst, and the rate (in diethyl ether) is increased by acids in the order: $ZnCl_2$ < $HCl \ll HBr < BCl_3 < SnCl_4 < FeCl_3 < AlCl_3 < FeBr_3 < TiCl_4$. Since this reaction involves much electronic rearrangement, it is not surprising to find that particles classified together on an electronic basis (i.e., Lewis acids) will react as catalysts in a fashion similar to Brönsted acids. Reactions whose catalysis depends on the specific nature of the proton will not be subject to catalysis by Lewis acids, however.

It is generally found that the various Lewis acids and bases do not have set orders of acidity and basicity, respectively. The relative basicity or binding strength of an electron-pair donor depends on the nature of the Lewis acid and vice versa. For example, iodide ion, thiosulfate ion, and ammonia all have stronger equilibrium binding constants to silver ion than does hydroxide ion, even though of the four the latter ion is the strongest base to protons.

Table 2-4   *Strength of Lewis acid-base reactions[a]*

Lewis base	Lewis acid[b]			
	H^+	Cu^{++}	Pb^{++}	Ag^+
I^-	-9		3.2	17.5
Br^-	-6	0.4	2.8	12.5
Cl^-	-3	2.4	2.8	8.6
H_2O	0.0	0.0	0.0	0.0
NO_3^-	0.4		2.1	-0.4
$S_2O_3^-$	3.6			16.5
SO_4^-	3.7	8.2		3.7
CH_3COO^-	6.5	9.4	3.8	4.1
C_5H_5N	7.0	11.8		7.7
NH_3	11.2	19.0		10.7
OH^-	17.5	22.2	9.5	7.1

[a] Reactions for these constants are listed below; all bases (nucleophiles) are denoted by N in the equations.

$$H_3O^+ + N \rightleftharpoons HN^+ + H_2O$$
$$Cu(H_2O)_4^{++} + 4N \rightleftharpoons Cu(N)_4^{++} + 4H_2O$$
$$Pb(H_2O)^{++} + N \rightleftharpoons PbN^{++} + H_2O$$
$$Ag(H_2O)_2^+ + 2N \rightleftharpoons AgN_2^+ + 2H_2O$$

[b] All values are log values relative to that for water.

Because it is not possible to set up a single scale of strengths for Lewis acids, the main employment of the Lewis theory has been in qualitative discussions of reaction rates and equilibria. By way of contrast, the Brönsted theory has been very useful quantitatively, even though more limited in scope.

In Table 2-4, some data related to binding strength in the Lewis sense are presented. It is apparent from these data that the dominating factors in determining binding strength to protons are not the same as those governing binding to some other Lewis acids. Indeed, any attempt to set up quantitative correlations between constants for rates and equilibria in Lewis acid-base reactions will of necessity invoke at least three contributing factors.

2-7 OTHER DEFINITIONS

The Brönsted and Lewis theories are classification schemes; they treat compounds and reactions in terms of the words *acid*, *base*, *neutralization*, *amphoterism*, etc. The same reactions and compounds have also been classified using different terminologies, including some words more commonly employed in the study of reaction kinetics and mechanism.

About 1927, Sidgwick[12] introduced the words *donor* and *acceptor* for electron-pair donor and electron-pair acceptor, respectively. The chemical combination of the donor and the acceptor gives a coordinate covalent bond. For example, NH_3 is a donor, Co^{3+} is an acceptor, and the complex formed is the hexamminecobaltic ion $[Co(NH_3)_6]^{3+}$. There is certainly a conceptual similarity between the ideas of Sidgwick and those of Lewis.

The physical organic chemists[13] employ the words *nucleophile* and *electrophile* for the donor and the acceptor, respectively. Nucleophile means "nucleus seeker," explaining what the non-bonded pair of electrons does in a chemical reaction; electrophile means "electron seeker." Most often, these two terms are used in connection with rates of displacement reactions, as will be seen in Chapter 4.

It is apparent that the problem of classification of chemicals and of reactions is partially one of semantics and partially one of personal choice. The hydroxide ion may be called a Brönsted base, a Lewis base, a donor, or a nucleophile; the terminology used is less important than the understanding gained of the chemical behavior of the hydroxide ion.

2-8 GENERAL INFLUENCE OF SOLVENTS

The medium in which a reaction takes place has an effect on the kinetics and mechanism. In view of the importance of medium effects, it is worth looking at some generalizations relative to them.

When a reaction can occur in both the gas phase and the liquid phase and when the kinetics are the same in both phases, the rate constants are usually quite comparable in size. An example is the first-order decomposition of dinitrogen pentoxide, for which data are presented in Table 2-5. The rate constants and activation parameters are similar for the reaction, whether in the gas phase or in a variety of solvents. Even though the mechanism is known to be complex (see Chapter 9), these solvents do not influence the general behavior.

On the other hand, many reactions that do not proceed in the gas phase are feasible in solvent systems. Solvents, particularly those of high dielectric constant, allow the intrusion of polar

Table 2-5 *The decomposition of dinitrogen pentoxide in various media*

Medium	$k \times 10^5$, sec^{-1}, 20°C	E_A, kcal mole^{-1}	$\ln_e A$	Note
Gas	1.65	24.7	31.48	a
N_2O_4	3.44	25.0	32.82	b
CH_3CHCl_2	3.22	24.9	32.56	b
$CHCl_3$	2.74	24.6	31.90	b
$CHCl_3$	2.14	24.5	31.22	c
$C_2H_4Cl_2$	2.38	24.4	31.42	b
CCl_4	2.35	24.2	31.05	b
CCl_4	2.34	24.5	33.09	c
$CHCl_2CCl_3$	2.20	25.0	32.35	b
Br_2	2.15	24.0	30.61	b
CH_3NO_2	1.67	24.5	31.13	b
$C_3H_6Cl_2$	0.24	28.3	35.72	b
HNO_3	0.05	28.3	34.11	b

[a] F. Daniels and E. H. Johnston, *J. Am. Chem. Soc.*, **43**, 53 (1921); E. C. White and R. C. Tolman, *J. Am. Chem. Soc.*, **47**, 1240 (1925).

[b] H. Eyring and F. Daniels, *J. Am. Chem. Soc.*, **52**, 1473 (1930).

[c] R. H. Leuck, *J. Am. Chem. Soc.*, **44**, 757 (1922).

intermediates and polar activated complexes. In such cases, the kinetics and the rate constants in solution bear little resemblance to those that would be predicted for the gas phase.

2-9 PHYSICAL PROPERTIES OF SOLVENTS

The rate of a chemical reaction can be a sensitive function of the nature of the solvent in which it is carried out. Because the interaction between the solvent and the solute (reactants, activated complex, products) can be chemically specific while depending at the same time on general physical properties, some of the factors important in solubility and in mechanism will be briefly mentioned. The correlation between solubility and mechanism is made here because high solubility implies a strong solvent-solute interaction, whereas a high rate implies a strong interaction of solvent and transition state.

Many reactions involve mechanistic steps in which particles undergo heterolytic scission to form ions. The ability of a solvent to aid the separation of a particle into ions depends on the dielectric constant ϵ. This constant is a measure of the solvent's ability to shield electric charges from each other. Coulomb's law is represented

$$\Delta F_e = -\frac{1}{\epsilon}\left(\frac{Z_C Z_A}{R_C + R_A}\right) \tag{2-12}$$

where ΔF_e is the electric free energy required to separate to infinity two ions of charge Z_A and Z_C and radius R_A and R_C, respectively. The dielectric constant of a vacuum is assigned the value 1; thus values of ϵ for various solvents are relative to that for a vacuum. Considering electrostatics alone, it can be seen that the separation of a molecule into ions is an endothermic process whose energy depends strongly on the dielectric constant. In Table 2-6, physical properties of six solvents are presented. The high dielectric constant of water lowers the free energy of charge separation to the extent that chemical solvation of the ions (always exothermic) makes the overall ionization process go to completion in most cases. In methanol, ion pairing is present; in sulfur dioxide and in dioxane, which have small dielectric constants, there are few free ions since the coulomb energy is so large that solvation energy cannot overcome this endothermicity.

Table 2-6 *Physical properties of six solvents*

Property[a]	H_2O	NH_3	HF	CH_3OH	SO_2	Dioxane
1. m.p.	0.0	-78	-83	-98	-75	11.7
2. b.p.	100.0	-33	19.5	64.6	-10	101.5
3. $\epsilon(T°C)$	84(0)	22(-34)	84(0)	35(13)	14(15)	2.2(25)
4. r	1.85	1.47	1.9	1.68	1.61	0^b
5. κ	5.5×10^{-8}	5×10^{-11}	1×10^{-5}	1×10^{-9}	4×10^{-8}	$<10^{-13}$
6. pK_s	14.0	32.7		17.0		
7. V_m, cc mole^{-1}	19	25	21	41	44	86

[a] Identified as follows: (1) melting point; (2) boiling point; (3) dielectric constant, with temperature of measurement in parentheses; (4) dipole moment in Debye units; (5) conductivity of pure solvent; (6) negative log of solvent-ion product; (7) molar volume.

[b] Although dioxane has a *zero* dipole moment for symmetry reasons, there is a strong quadrupole moment that is important in solvation.

The dipole moment r of the solvent molecule also stabilizes ions, one contribution to specific solvent-solute interaction being the ion-dipole energy expressed by the equation

$$\Delta F = -\frac{Zr}{R^2} \tag{2-13}$$

where R is the distance and ϵ is assumed to be 1. In similar fashion, the polarizability α of the solvent molecule contributes by ion-induced dipole attraction in accord with the equation

$$\Delta F = -\frac{Z^2\alpha}{R^4} \tag{2-14}$$

with other symbols the same as for Eq. (2-13). The interaction of an ion with a polar and/or polarizable molecule is always favorable from an energy standpoint; thus the endothermicity from separation of a particle into ions is often compensated for by the exothermic interaction of the ions with solvent molecules.

Other physical properties of the solvent that may be of interest in the study of mechanism are: viscosity; liquid range (melting point, boiling point, and the convenience of their measurement); and molar volume (since specific solvation is easier with a smaller solvent molecule).

2-10 CHEMICAL NATURE OF SOLVENTS

The acidic or basic nature of the solvent can be a dominant factor in rates and mechanism. For example, certain peroxide reactions whose rates are too slow to be measured in water solution have rates that can be followed in solvents such as acetic acid or sulfuric acid. There are also basic solvents such as liquid ammonia in which the behavior of solvated electrons can be investigated.†

The addition of acids to a basic solvent such as liquid ammonia results in the formation of the solvated proton NH_4^+; this is the strongest acid that can exist in this solvent. In similar fashion, the strongest base that can exist in acetic acid is the acetate ion. Generally, the strongest acid that can exist in any solvent system is the conjugate acid (*lyonium ion*) of the solvent molecule, and the strongest base that can exist is the conjugate base (*lyate ion*) of the solvent molecule. The ability of the solvent system to lower the acidity and basicity to that of the lyonium and lyate ions respectively, is called the *leveling effect* of the solvent.

The range of acid strengths available for reaction catalysis is dependent on both the acidity and the basicity of the solvent. A solvent that is both strongly acidic and basic will have a high equilibrium self-ionization.

$$2HS \rightleftharpoons S^- + H_2S^+$$

The amount of self-ionization is evaluated from the expression

$$K_s = [S^-][H_2S^+] \tag{2-15}$$

and is usually recorded as pK_s, i.e., $-\log K_s$. A low pK_s indicates high self-ionization and a low range of available acidities to added solutes; one such strongly leveling solvent is formic acid. Conversely, a solvent like nitrobenzene that is neither acidic nor basic will have a wide range of possible solute acidities for the running of reactions.

Hydrogen bonding between the solute and the solvent is a chemical reaction directly related to the rapid reactions of acids and bases. Transfer of hydrogen ions can be accomplished by a path of low activation energy involving hydrogen-bonded structures as

† A great deal of work has been carried out on reactions in nonaqueous solvent systems, and there is even a *solvent theory of acids and bases*. Data on these systems and on the theory may be found in many inorganic texts (cf. Ref. 4).

intermediates. This is presumably the reason why O—H and N—H ionization reactions are much faster than C—H reactions even though the former bonds have higher energies. A specific example of hydrogen bonding between solvents and other particles in an activated complex will be pointed out in Chapter 5. Hydrogen bonding can also affect the stability and kinetic reactivity of ions (reactants, transition states, products) in a particular solvent, a fact that has recently been emphasized by Parker.[15]

A nonbonded pair of electrons on a solvent molecule is a site for chemical attachment of the solvent molecule to a solute. As will be seen in the sections on the behavior of coordination compounds, such spare pairs of electrons make the solvent a ligand; this allows a mechanistic path in which the solvent molecule replaces one ligand and is in turn replaced by a second.

The problem of solvent effects on rates and mechanisms is very important and has been neglected or oversimplified in many studies. Since it is not possible to cover this topic in detail here, the reader is referred to the kinetics and mechanisms books listed at the end of Chapter 1 and to the interesting papers in the recent literature.[14, 15]

References

1. L. P. Hammett, *Physical Organic Chemistry*, McGraw-Hill, New York, 1940, Chap. VII.
2. J. N. Brönsted, *Chem. Rev.*, 5, 322 (1928).
3. J. F. Bunnett, "Investigation of Rates and Mechanisms of Reactions," in S. L. Friess, E. S. Lewis, and A. Weissberger (eds.), *Technique of Organic Chemistry*, 2d ed. (revised), Interscience, New York, 1961, Vol. VIII, Part 1.
4. R. P. Bell, *Acids and Bases—Their Quantitative Behavior*, Methuen, London, 1952; T. Moeller, *Inorganic Chemistry*, Wiley, New York, 1952, p. 306.
5. J. N. Brönsted, *Rec. Trav. Chim.*, 42, 718 (1923).
6. R. P. Bell, *Acid-Base Catalysis*, Clarendon Press, Oxford, 1941.
7. See R. P. Bell, *Acids and Bases—Their Quantitative Behavior*, Methuen, London, 1952, p. 4.
8. J. N. Brönsted and K. Pedersen, *Z. Physik. Chem. (Leipzig)*, 108, 185 (1924).
9. S. W. Benson, *J. Am. Chem. Soc.*, 80, 5151 (1958).
10. G. N. Lewis, *Valence and the Structure of Atoms and Molecules*, Chemical Catalog Co., New York, 1923; see also G. N. Lewis, *J. Franklin Inst.*, 226, 293 (1938); W. F. Luder and S. Zuffanti, *The Electronic Theory of Acids and Bases*, Wiley, New York, 1946.
11. R. P. Bell and B. G. Skinner, *J. Chem. Soc.*, 1952, 2955.

12. N. V. Sidgwick, *The Electronic Theory of Valency*, Clarendon Press, Oxford, 1927.
13. Cf. C. K. Ingold, *Structure and Mechanism in Organic Chemistry*, Cornell University Press, Ithaca, N.Y., 1953, p. 200 ff.
14. Cf. E. A. S. Cavell, *J. Chem. Soc.*, **1958**, 4217; S. Winstein et al., *Tetrahedron Letters*, **1960** (No. 9), 24; also R. G. Pearson and D. C. Vogelsong, *J. Am. Chem. Soc.*, **80**, 1048 (1958); J. Miller and A. J. Parker, *J. Am. Chem. Soc.*, **83**, 117 (1961).
15. A. J. Parker, *J. Chem. Soc.*, **1961**, 1328; A. J. Parker, *Quart. Rev. (London)*, **16**, 163 (1962).

3

Binding Strength and Reaction Rates: Linear Free-Energy Relations

The *law of microscopic reversibility* (Section 1-4) defines the relationship between rate constants and equilibrium constants. From the form of the equation for microscopic reversibility, [Eq. (1-6)], it can be seen that the equilibrium constant determines only a ratio of rate constants and does not give any information as to their absolute values. To effect an understanding of the relation between rate and equilibrium constants for specific cases, it has been necessary first to correlate empirically the observed constants and then to explain the results. Such empirical correlations have been carried out through linear free-energy relations,[1-6] through postulation and search for intermediates, and through kinetic isotope effects.

These three approaches, which are the topics included in this chapter, are of importance to the study of mechanisms because the information they yield relates to such questions as the extent of bond formation and bond breaking in the transition state; the type of bonding present in the activated complex; the directions and magnitudes of electronic influences on rate; and identification of intermediate species present in the detailed steps of the reaction mechanism.

3-1 LINEAR FREE-ENERGY RELATIONS

Consider the reaction

$$A_n + B\text{---}C_m \to A_n\text{---}B + C_m$$

where A_n is one of a closely related series of compounds, C_m is one of another series, and the activated complex is of the type (I). The bond $A_n\text{---}B$ is being formed in the transition state while the $B\text{---}C_m$ bond is breaking; thus the reaction-coordinate diagram has only a single maximum near the center.

$$A_n \text{ - - - } B \text{ - - - } C_m$$

(I)

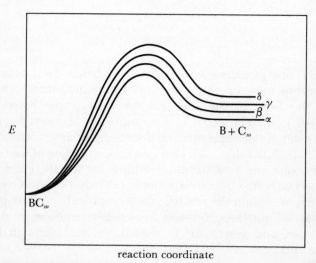

Figure 3-1 *Hypothetical reaction-coordinate diagram for the reaction $B\text{---}C_m \to B + C_m$ (or for $A + B\text{---}C_m \to A\text{---}B + C_m$) wherein the strength of the $B\text{---}C_m$ bond increases from α to δ.*

Looking first at the B—C_m bond, it is apparent that the activation energy should reflect in some measure the bond strength since the endothermicity that results from breaking this bond contributes to the activation energy. The reaction-coordinate diagram for the case where bond breaking is more important than bond making will take the form of Figure 3-1, with the bond strength increasing from α to δ. In the extreme case where bond breaking is complete and there is no bond formation, the activation energies will directly parallel the endothermicities.

Since B becomes partially available for bonding to A_n as the reaction proceeds, there is in the transition state a certain amount of strength in the new bond. This strength should be related to the ability of A_n to form a stable bond in the product. Figure 3-2 shows curves for the case where bond making is more important than bond breaking; in this reaction-coordinate diagram, the A_n—B bond

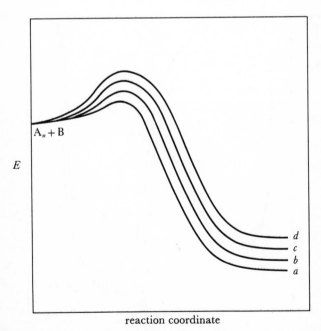

reaction coordinate

Figure 3-2 *Hypothetical reaction-coordinate diagram for the reaction $A_n + B \rightarrow A_n$—B (or for $A_n + B$—$C \rightarrow A_n$—$B + C$) wherein the strength of the A_n—B bond increases from **d** to **a**.*

strength increases as one goes from d to a. It is seen that the most stable product tends to have the largest effect in lowering the activation-energy hill; in general, the activation energy decreases as the stability of the product increases.

Bell[3] has shown, using Morse curves, that changes in activation energies are proportional to changes in ground-state energies for products of related reactions. In Figure 3-3, the energy profiles for the transfer of B from reactant BC to the two product compounds BA_n and BA_{n+1} are shown. The Morse curves for BA_n and BA_{n+1} are drawn with a constant difference in E; they have the same shapes but differ in values of E_0 (the ground-state energy). The activation energy for either of the two cases is the vertical difference between $E_0(BC)$ and the point where the BC curve crosses

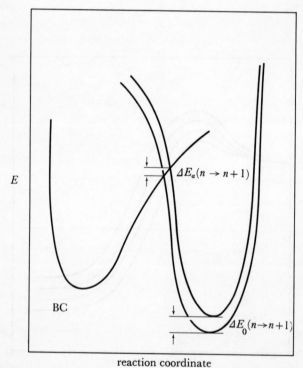

reaction coordinate

Figure 3-3 *Hypothetical reaction-coordinate diagram based on Morse potential curves for the reaction $A_n + B{-}C \to A_n{-}B + C$. The relation between activation-energy differences and product-bond-energy differences is shown.*

the appropriate AB curve. From the geometry of the system, it can be shown that

$$\Delta E_a(n \to n + 1) = \beta \Delta E_0(n \to n + 1) \tag{3-1}$$

where ΔE_a is the activation-energy difference, ΔE_0 is the ground-state-energy difference, and β is a constant characteristic of the reaction series. In practice, β takes values between 0 and 1. If the entropy contributions are constant, then it follows that

$$\Delta[\Delta F^{\ddagger}(1 \to 2)] = \beta \Delta[\Delta F(1 \to 2)] \tag{3-2}$$

where $\Delta[\Delta F^{\ddagger}(1 \to 2)]$ is the difference in activation free energies for two cases and where $\Delta[\Delta F(1 \to 2)]$ is the difference in free energies of reaction for the same cases.

Eq. (3-1) relates the energy of activation to the energy of the bond that is being formed. On the basis of microscopic reversibility, it is equally true that for a symmetrical activated complex the energy of activation is related to the absolute value of the energy of the bond that is being broken. Using previous arguments, one can derive an analogous equation

$$\Delta E_a(m \to m + 1) = \beta' \Delta E(m \to m + 1) \tag{3-3}$$

for comparison of energies of activation with strengths of reactant bonds. Again, the assumption that entropies are constant leads to an equation relating free energies of activation to free energies of reaction.

These free-energy equations can be rewritten in terms of rate constants or equilibrium constants; for example,

$$\log (k_1/k_2) = \beta \log (K_1/K_2) \tag{3-4}$$

This is one of a general group of equations called *linear free-energy equations*. As a group, they are examples of what is known as the *principle of linear free-energy relations*. In words, this principle states that the free-energy changes for one series of reactions are linearly related to those for a similar series.[†] The symbol LFER will be used in this book to stand for linear free-energy relation(s). There are a variety of possible versions of linear free-energy equations. For example, one can correlate rates with equilibria, rates with rates,

[†]A more formal treatment of linear free-energy relations may be found in the chapter entitled, "Theoretical Introduction to Extrathermodynamic Relationships," in J. E. Leffler and E. Grunwald, *Rates and Equilibria of Organic Reactions*, Wiley, New York, 1963.

equilibria with equilibria. Application of these equations will be made in Section 3-4.

3-2 ASSUMPTIONS

In the presentation above there are two basic assumptions. The first is that the Morse curves are parallel for related compounds; the second is that free energies can be directly related to heats of reaction (that is, that the entropy changes accompanying the formations of B—A_n and B—A_{n+1} are the same). Both of these assumptions deserve further comment.

The condition that the Morse curves for similar reactants be parallel (differ by a constant amount of energy at all values of internuclear distance) is met only by a fortuitous set of circumstances over a limited range of internuclear separations. A more realistic approximation is that, as the internuclear distance increases above the equilibrium value, the energy separation of the Morse curves becomes a decreasing fraction of the ground-state-energy difference. The use of this latter approximation will not alter the conclusion that activation-energy changes are a function of ground-state-energy changes. This approximation is also more consistent with the present usage of reaction-coordinate diagrams.

The term *reaction series* finds its meaning in the Morse curves and their application to the discussion of reaction-coordinate diagrams. From this point of view, a reaction series is a group of compounds with similar potential curves and therefore similar reaction-coordinate diagrams. In the Brönsted equation, the reaction series is a group of compounds having the general formula ROH; variations in R cause changes in the ground-state energy of the oxygen-hydrogen bond and cause related changes in the activation energy for breaking this bond in the transition state. In general, the assumption of similar Morse curves is reasonable only when the group being varied in the reaction series is not adjacent to or strongly bonded by resonance to the atoms whose bonds are being formed or broken. From the general chemist's point of view, a reaction series is a group of compounds that can be expected to show similar behavior.

The assumption of constant entropy changes is, unfortunately, often incorrect. In many reaction series, the values of ΔS vary markedly for the individual reactants. However, this entropy variation is generally accompanied by a compensating enthalpy variation; thus the free-energy changes are linearly related even though the heats and entropies are not simply related. This point will be more fully discussed in Section 3-5 of this chapter.

3-3 MATHEMATICAL FORM

The equations relating equilibrium constants and rate constants to their respective free energies are

$$\Delta F = -RT \ln K \tag{3-5}$$

and

$$\Delta F^{\ddagger} = -RT \ln k + RT \ln \left(\frac{kT}{h}\right) \tag{3-6}$$

Thus, any change in free energy appears as a logarithmic function of the measured constant. For this reason, all LFER are treated in terms of the mathematical form

$$\log K_x = G + \alpha \log K'_x \tag{3-7}$$

where K_x and K'_x are equilibrium and/or rate constants involving the same reactant in different reactions. Usually a graph is made of the data, with each point on the plot representing the two constants for one reactant. The quantities G and α are intercept and slope, respectively, for the plot of $\log K_x$ against $\log K'_x$. The slope α, which will be termed the *substrate constant*, is the value of greatest interest; it can give information on the reaction mechanism that cannot be gained from a study of the rate law and the rate constant for a single reactant. In the Brönsted equation, for example, the value of the substrate constant gives a measure of the amount of bond breaking in a transition state as compared to the amount of bond breaking at equilibrium (as denoted by the acid-ionization constant). The value of the intercept G is dependent on many nonrelative (i.e., absolute rate constant) and external (i.e., ionic strength) factors and is, therefore, of only limited usefulness.

3-4 EXAMPLES OF LFER

The Brönsted equation[1,3,4] for correlating rate constants in a general-acid- or general-base-catalyzed reaction with the corresponding equilibrium acid- or base-ionization constants was the earliest LFER. The form of this equation [cf. Eqs. (2-10) and

(2-11)], along with some experimental data (Table 2-2) and a resultant plot (Figure 2-1), were presented in Chapter 2.

In recent years, it has become apparent that Brönsted-type plots are observed when *nucleophilic catalysis* rather than general-base catalysis is the basis of the correlation. The mechanism for nucleophilic catalysis of a hydrolysis reaction may be schematically described as

$$B + X—Y \rightarrow B—X^+ + Y^-$$

$$H_2O + B—X^+ \rightarrow HO—X + HB^+$$

where the first step is rate-determining, and where B is a base and X—Y a substrate. Nucleophilic catalysis has been observed in the hydrolysis of reactive esters such as *p*-nitrophenyl acetate[12, 13] and in the hydrolyses of the fluorophosphates.[14]

One of the most useful of the LFER is the Hammett equation,[2, 4, 5, 6] which is often written

$$\log (k_x/k_0) = \sigma_x \rho \tag{3-8}$$

where the term (k_x/k_0) is a ratio of the constant for a *para-* or *meta*-substituted benzene compound to that for the unsubstituted compound, ρ is a substrate constant, and σ_x is the logarithm of the ratio of the ionization constant for the substituted benzoic acid to that for benzoic acid itself. Eq. (3-8) was derived from the relationship

$$\log k_x - \log k_0 = \rho(\log K_x - \log K_0) \tag{3-9}$$

which fits the general form of LFER equations since the group $[\log k_0 - \rho (\log K_0)]$ is an intercept (G) for any series of reactants. The substrate constant ρ is a measure of the influence of a substituent group on a specified reaction of a benzene compound as compared to its influence on the ionization of benzoic acid. The Hammett equation has been particularly useful for explaining electronic effects on equilibria and on rates and mechanisms of reactions of aromatic organic compounds.

Both Brönsted[1] and Hammett[2] used pK_a values for proton acids as constants of the standard reaction series. Because of the large number of precise pK_a values available, they form an excellent standard set of constants. However, the rates of many reactions do not correlate well with the reactant acidities (or basicities); therefore it is necessary to define other standard reaction series for such cases. Swain and Scott,[15] in a survey of the reactions of nucleophiles with saturated carbon compounds, showed that the order of strength of various nucleophiles seemed to be fairly independent of the substrate. They proposed the equation

$$\log (k_n/k_0) = sn \tag{3-10}$$

where k_n is the rate of reaction of nucleophile N with a carbon compound, k_0 is the rate of reaction of a water molecule with the carbon compound, s is the substrate constant, and n is the nucleophilic constant. This constant n is obtained from the relative rates of attack of the two nucleophiles, N and water, on the standard substrate, methyl bromide. Certain of its values are listed in Table 4-1, Chapter 4.

Swain and Scott[15] pointed out that their constants would not work with substrates other than carbon compounds, and it was suggested that separate scales be set up to correlate rates of nucleophilic attack on such other substrates as hydrogen and sulfur. As will be discussed in Chapter 4, many inorganic substrates do react with the nucleophiles and show different orders of reactivity. The orders are, however, not independent of each other. On the assumption that the several orders are related, it was proposed[16] that nucleophilic character is a combination of several factors; each of these factors contributes to a greater or lesser extent in the reaction with a particular substrate. The equation that has been suggested is

$$\log (k_n/k_0) = \alpha E_n + \beta H \tag{3-11}$$

where k_n and k_0 are defined as in Eq. (3-10), α and β are two substrate constants, and E_n and H are nucleophilic constants. The nucleophilic constants E_n and H are derived from electrode-potential data and proton basicities, respectively. Eq. (3-11), along with a similar one[17] relating nucleophilic strength to basicity and polarizability, have been employed to correlate a large amount of rate and equilibrium data for a variety of systems. Their use in rate correlations will be discussed in Chapter 4.

Of somewhat different nature is the Grunwald-Winstein relation,[4, 18] which deals with the solvent effect on rates of reactions involving charge separation (ionization and electrolytic-dissociation processes). For the standard reaction, they used the solvolysis of tert-butyl chloride

$$(CH_3)_3CCl \rightarrow (CH_3)_3C^+ + Cl^-$$

$$(CH_3)_3C^+ + HS \rightarrow \text{products}$$

where the first step is rate-determining and where HS denotes solvent. They devised the relation

$$\log k - \log k_0 = mY \tag{3-12}$$

In this equation, the rate constant k for a reaction in any solvent is compared with the constant k_0 for the standard solvent (80 per cent ethanol, 20 per cent water). The substrate constant m is a measure of the sensitivity of the particular reaction to the ionizing power of the solvent as compared to the sensitivity of *tert*-butyl chloride. The parameter Y is defined by the equation

$$Y = (\log k - \log k_0)_{tert-\text{BuCl}} \tag{3-13}$$

and is a measure of the relative ionizing powers of various solvent systems as determined by their rates of solvolysis of *tert*-butyl chloride. Values of Y have been reported for a number of solvent systems. The Grunwald-Winstein relation has made it possible to correlate the rates of a variety of reactions such as solvolyses and rearrangements in which ionization occurs in the rate step. The empirical value of Y seems to be a better measure of the ionizing power of the solvent than is the dielectric constant; at least in part, this is because the solvent effect on a transition state is much more specific than can be judged from the macroscopic dielectric constant alone.

In principle, it is possible by using LFER to sort out the various factors that contribute to the rate of a reaction. Taft[6] has written a long article summarizing his and other workers' efforts to separate polar, steric, and resonance effects on reactivity. Numerical constants for the inductive effect, resonance effect, and even steric size of a group are presented, along with rational grounds for evaluation and application of these constants. The results are encouraging and interesting, and the reader will profit by careful study of the Taft approach.

3-5 PROBLEMS WITH LFER

The employment of these log-log plots, which are fundamentally empirical approaches toward understanding rates and equilibria, has multiplied in recent years. Their application is helpful in mechanism studies; yet their usefulness may be greater when the agreement is poor rather than good. For example, the fact that bond breaking is occurring in a transition state can be shown clearly by a plot of the log of the rate constant against the log of the basicity of the leaving anion. This fact is not likely to surprise an investigator, since the rate law and the first two constants measured would quickly lead him to expect it. He might be surprised, however, by the possible presence of one point that deviates markedly from an otherwise excellent plot. Often the nature of the discrepancy is something scientifically new. For this and other reasons, it is worth looking at some of the problems that arise in the employment of these free-energy relations.

One problem that has been recognized for some years and has been reviewed recently[4,19] is that though the free energies may correlate well (form a linear plot), the enthalpies and entropies may behave in an erratic manner. The problem can be summarized in the following way: Despite the fact that derivations of the LFER principle require the assumption that entropy changes for a reaction series be constant (so that free energies can be directly related to enthalpies), good linear log-log plots are obtained even when the entropies behave in an irregular fashion.

For two reactions with similar free energies, the only restriction on the values of ΔS and ΔH is expressed by the equation

$$\Delta H_1 - T(\Delta S_1) = \Delta H_2 - T(\Delta S_2)$$

or

$$\Delta H_1 - \Delta H_2 = T(\Delta S_1 - \Delta S_2) \tag{3-14}$$

where the subscripts 1 and 2 refer to the two reactions. Thus it is possible to have small and predictable changes in ΔF for the compounds in a reaction series along with large variations in ΔH and ΔS; however, the changes in ΔH must parallel the changes in ΔS.†

A reaction in which changes in substituent have little effect on $\Delta F^{\ddagger}$ but significant effect on $\Delta H^{\ddagger}$ and $\Delta S^{\ddagger}$ has been reported by Fraser.[20] The general equation for this reaction is

$$(NH_3)_5CoL^{++} + 5H^+ + Cr^{++} \rightarrow 5NH_4^+ + Co^{++} + CrL^{++}$$

---

† One source of parallel deviations in ΔH and ΔS is experimental error. Even though rate constants may be known to within ± 5 per cent in a particular case, so that $\Delta F^{\ddagger}$ values are good to within 0.1 kcal mole^{-1}, the values of $\Delta H^{\ddagger}$ and $\Delta S^{\ddagger}$ can be off by as much as 1 kcal mole^{-1} and 5 cal mole^{-1} deg^{-1}, respectively, when data are taken at only two temperatures 10° apart.

where L is a carboxylato ligand. The rate law for this oxidation-reduction reaction involving ligand transfer is

$$v = k[(NH_3)_5CoL^{++}][Cr^{++}] \qquad (3\text{-}15)$$

and the activated complex presumably has the structure (II).

$$\left[\begin{array}{cccc} H_3N & NH_3 & H_2O & OH_2 \\ & \diagdown\diagup & & \diagdown\diagup \\ H_3N-Co-O & & O-Cr-OH_2 \\ & \diagup\diagdown & C & \diagup\diagdown \\ H_3N & NH_3 & R\ H_2O & OH_2 \end{array} \right]^{4+}$$

(II)

Data for rates with a variety of ligands are presented in Table 3-1. The small variations in k (and thereby in $\Delta F^{\ddagger}$) are to be contrasted with the sizable and congruent variations in $\Delta H^{\ddagger}$ and $\Delta S^{\ddagger}$. For purposes of comparison, it is worth noting that a shift of either 1.4 kcal mole^{-1} in $\Delta H^{\ddagger}$ or 5 cal mole^{-1} deg^{-1} in $\Delta S^{\ddagger}$ alone will shift a rate constant by one order of magnitude.

A second problem arises in the occasional observation that a series of rates does not parallel the thermodynamic driving force.

Table 3-1 *Activation parameters for chromous ion and pentammine carboxylatocobalt*(III) *ion reactions*[a]

Ligand	k	$\Delta H^{\ddagger}$	$\Delta S^{\ddagger}$
Acetato	0.18	3.5	−50
Chloroacetato	0.10	7.9	−37
Cyanoacetato	0.11	4.0	−49
Dichloroacetato	0.074	2.5	−55
Benzoato	0.14	4.9	−46
o-Chlorobenzoato	0.074	6.0	−43
o-Iodobenzoato	0.082	2.8	−54
o-Phthalato	0.075	5.1	−45
m-Phthalato	0.093	2.6	−56
p-Chlorobenzoato	0.21	10.0	−28
p-Hydroxybenzoato	0.13	9.6	−30
p-Cyanobenzoato	0.18	7.5	−37
p-Sulfobenzoato	0.16	8.3	−34

[a] See text for stoichiometry and rate law.

This situation is permitted by the law of microscopic reversibility, which puts no restrictions on the magnitude of k_f or k_r as long as their ratio remains equal to the equilibrium constant. An interesting treatment of this problem has been given by Hammond[21] who has proposed a simple postulation to aid in estimating the degree of correlation between rate constant and equilibrium constant that can be expected in a given reaction. The postulate is: "If two states, as, for example, a transition state and an unstable intermediate, occur consecutively during a reaction process and have nearly the same energy content, their interconversion will involve only a small reorganization of the molecular structures."

A reaction in which the rates of product formation have little correlation with product stabilities is the formation of glycol complexes of tellurate ion.[22].

The rate law is

$$v = k[\text{H}_5\text{TeO}_6^-][\text{OH}^-][\text{glycol}] \tag{3-16}$$

and the values of the rate constant k vary little with the nature of the glycol. The suggested mechanism includes TeO_4^- as a high-energy intermediate before the rate-determining step. From this hypothesis and the Hammond postulate, it follows that there should be only loose association of TeO_4^- and glycol in the transition state and that the rates should depend little on the equilibrium binding strength of the final complex; this is just what is observed in the reaction. The necessary consequence in this case, that the binding strengths of the individual complexes show up in the reverse-rate constants, has also been observed.

Hammond's treatment is useful for those mechanisms that involve intermediates and highly exothermic or endothermic steps. There are, however, some cases without these phenomena in which rate constants when graphed against equilibrium constants do not form a good linear plot. For example, it is known[13, 23] that some nucleophiles show reactivity far beyond what one would expect from their binding strengths. Presumably, such reactivity results from exceptional binding in the transition state. Since the nature of such binding will be specific for each chemical case, it will not be discussed here. It is appropriate, however, to consider briefly the consequence of an exceptional kinetic reactivity.

Figure 3-4 presents two pertinent cases on one reaction-coordinate diagram. Both curves begin and end at the same energies, and their only differences lie in the area of the transition state. It is apparent there that a low activation energy and high reactivity in the forward direction must be accompanied by a low activation energy and high reactivity in the reverse direction. It follows, then, that a donor with high nucleophilic reactivity is easily displaced from its compounds. One such case is that of the peroxyanions ROO^-, which are known to be excellent nucleophiles.[13,23] Since they do not seem to form exceptionally strong bonds, it can be predicted that they should be good *leaving groups*. Indeed it has been found that peroxyphosphates are remarkably susceptible to hydrolysis by a displacement mechanism, with the peroxyanion the leaving group.

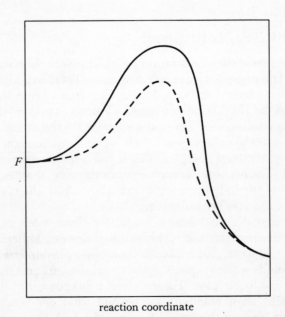

reaction coordinate

Figure 3-4   *Reaction-coordinate diagram showing how two similar reactions of the same energy may have quite different activation energies.*

Another problem arises in the separation of the over-all free energy into individual contributions (bond making, bond breaking, etc.) and into external factors (ionic strength, temperature, etc.), all of which can be demonstrated by some linear plot. Miller[24] has pointed out that the usual procedure is based on the implicit assumption of independent contributions. This assumption is not always justified since there will be *cross-terms*; that is, a change in one variable will alter the magnitude of the response to change in some other variable. For the reaction of A_n with $B—C_m$, the slope of the log-log plot of rate against basicity of C_m will depend on the A_n species employed. In terms of free energies,

$$\Delta[\Delta F \text{ (over-all)}] = \Delta[\Delta F(A_n)] + \Delta[\Delta F(C_m)] + \Delta[\Delta F(A \times C)]$$

$$(3\text{-}17)$$

where the four $\Delta(\Delta F)$ values represent, respectively, the over-all change, the change resulting from A_n variation, the change resulting from C_m variation, and the change resulting from the influence of A_n and C_m on each other.

3-6 INTERMEDIATES

Because intermediates occur in most reaction sequences, the nature of these transient chemical species and the possible ways of establishing their presence in a mechanism will be brought out here. An intermediate is a molecular configuration passed through during the course of a reaction sequence and existing long enough to exhibit definite chemical properties such as selectivity. As mentioned in Chapter 1, intermediates are denoted by minima in a reaction-coordinate diagram.

The best evidence for an intermediate is isolation and analysis, although intermediates by their very nature are often too reactive to be isolated. An example of an achieved isolation is the identification and separation of hydroperoxides from autoxidation reactions. Physical methods that are based on some property of the intermediate are useful in demonstrating their presence; spectroscopy has been especially valuable in this regard. Another way of detecting intermediates is by *trapping* experiments. A substance that does not react either with starting material or products but probably will react with the intermediate is added to the reaction mixture; an adduct of the added substance and the intermediate, often can be isolated. By trapping it with trimethylamine, borane

(BH_3) has recently been shown[25] to be an intermediate in the hydrolysis of borohydride ion. Intermediates have also been shown to exist in reactions of the type

$A \rightarrow B$

$B \rightarrow C$

where chemical analyses for both A and C showed their sum at some intermediate time to account for less than the initial amount of A.

Kinetics studies often provide evidence for intermediates. Induction periods are caused by a slow build-up of an intermediate. If the orders in the rate law are less than the stoichiometric numbers, there is an intermediate formed in the rate-determining step.† The presence of an intermediate is also indicated if the rate law contains inverse orders, nonintegral orders, or orders larger than three (since molecularities higher than three are unlikely). Finally, the applicability of the Hammond postulate will suggest the presence of an intermediate in some cases.

3-7 KINETIC ISOTOPE EFFECTS

It is known that the isotopic mass of an atom at a reaction site has an influence on the rate. Therefore, it is possible to design isotope-labeling experiments to test a postulated mechanism in which a particular bond is broken in the transition state. The magnitude of the observed change in rate constant will depend on the ratio between the masses of the isotopes. Thus, larger effects will be observed with isotopes of a light atom than with those of a heavy atom; the sizes of the rate differences are usually predictable.

The fundamental consideration in the theory[7-11] treating kinetic isotope effects is that all chemical bonds have a *zero-point energy* and that this energy in the transition state will be different from this energy in the ground state, particularly if the bond is

†The existence of an intermediate in the oxidation of aniline to nitroso-benzene

$$2CH_3CO_3H + C_6H_5NH_2 \rightarrow 2CH_3CO_2H + C_6H_5NO + H_2O$$

is demanded by the observation that the rate law is first order in each reactant. It was also shown that the reaction (as followed by product nitrosobenzene formation) had an induction period. The intermediate is almost certainly N-phenylhydroxylamine [K. M. Ibne-Rasa and J. O. Edwards, *J. Am. Chem. Soc.*, **84**, 763 (1962)].

being broken. The zero-point energy E_0 of the bond is a residual energy, equal to the difference between the energy of the lowest real state and the minimum of the potential curve. This energy may be calculated from the equation $E_0 = \frac{1}{2}h\nu$, where ν is the bond-vibration frequency. For two isotopes in an otherwise identical bond, the difference between the two zero-point energies is given by

$$E_0 = \tfrac{1}{2}h(\nu - \nu') \qquad\qquad (3\text{-}18)$$

where ν and ν' refer to bonds containing the lighter and heavier isotopes X and X', respectively; the ratio of ν to ν' is inversely proportional to the square root of the ratio between the masses of the isotopes. In the case of the isotopes H and D, where a large relative mass difference exists, the value of ΔE_0 is 1.4 kcal mole^{-1} for bonds to oxygen. The bond with the lighter isotope has the larger zero-point energy and is, therefore, higher on the potential curve; as a result, it is less stable than the bond with the heavier isotope. Should there be complete loss of bonding to the hydrogen atom in the transition state, the values of $E_0^{\ddagger}$ (zero-point energy in the transition state) and $\Delta E_0^{\ddagger}$ will both be 0. It will therefore require more activation energy (1.4 kcal mole^{-1}) for the deuterium compound than for the protium compound to undergo reaction. From potential curves such as those in Section 3-1 on LFER, it can be seen that the activation energy for breaking any oxygen-deuterium bond will be greater than that for breaking an oxygen-protium bond as long as the value of $\Delta E_0^{\ddagger}$ is smaller than that of ΔE_0, as is shown in Figure 3-5. This condition seems a reasonable one when the bond is being broken in the rate step, for any weakening of a bond will lower the values of ν and ν' proportionally and thereby lower $\Delta E_0^{\ddagger}$.

With the assumption that $\Delta E_0^{\ddagger}$ is 0, it is possible to compute expected changes in rates for bond-breaking processes involving isotopes. A list of such predicted rate effects is presented in Table 3-2. These data show that kinetic isotope effects are small, even with elements as light as carbon, although striking rate changes occur with the hydrogen isotopes.

The bromination of nitromethane[26] follows the rate law

$$v = k[CH_3NO_2][B] \qquad\qquad (3\text{-}19)$$

where B is a general base. This suggests the mechanism

$$CH_3NO_2 + B \rightarrow CH_2NO_2^- + HB^+$$

$$CH_2NO_2^- + Br_2 \rightarrow BrCH_2NO_2 + Br^-$$

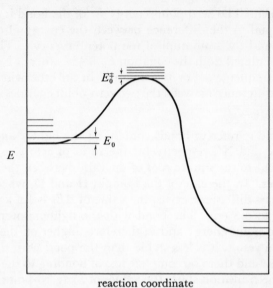

E

reaction coordinate

Figure 3-5 *Reaction-coordinate diagram showing the influence of the zero-point energy in reactants (E_0) and in transition state ($E_0^{\neq}$) on the free energy of activation.*

The first step, which must be rate-determining, involves a breaking of the carbon-hydrogen bond; with acetate ion acting as a general base, the value of 6.53 has been observed for k_H/k_D. This value

Table 3-2 *Predicted kinetic isotope effects*

Bonds		Rate factor[a]
C—H	C—D	6.9
N—H	N—D	8.5
O—H	O—D	10.6
H—H	H—D	3.7
H—H	D—D	19.5
C^{12}—C^{12}	C^{12}—C^{13}	1.07

[a] Defined as the rate constant for the light isotope divided by the rate constant for the heavy isotope.

is consistent with that predicted for a kinetic isotope effect and therefore supports the mechanistic postulation.

The technique of isotopic substitution at a reaction site is useful in the evaluation of mechanism. The results, however, must be subjected to careful scrutiny, for the following factors can influence the observed isotope effect. Constants for equilibria prior to the rate step may be subject to isotope effects: for example, the ionization constants of proton acids are significantly different, larger by a factor of approximately three in H_2O than in D_2O. Isotopic substitution at a nonreactive position can influence the rate, sometimes to the extent of overwhelming the expected primary influence.† Finally, every isotope study must be thoroughly checked to make sure that an exchange process does not vitiate the conclusions.

References

1. J. N. Brönsted, *Chem. Rev.*, **5**, 322 (1928).
2. L. P. Hammett, *Physical Organic Chemistry*, McGraw-Hill, New York, 1940, Chap. VII.
3. R. P. Bell, *The Proton in Chemistry*, Cornell University Press, Ithaca, N.Y., 1959, p. 124; R. P. Bell, *Acid-Base Catalysis*, Clarendon Press, Oxford, 1941.
4. J. F. Bunnett, "Investigation of Rates and Mechanisms of Reactions," in S. L. Friess, E. S. Lewis, and A. Weissberger (eds.), *Technique of Organic Chemistry*, 2d ed. (revised), Wiley-Interscience, New York, 1961, Vol. VIII, Part 1.
5. H. H. Jaffe, *Chem. Rev.*, **53**, 191 (1953).
6. R. W. Taft, Jr., in M. S. Newman (ed.), *Steric Effects in Organic Chemistry*, Wiley, New York, 1956, Chap. 13.
7. J. Bigeleisen and M. Wolfsberg, "Theoretical and Experimental Aspects of Isotope Effects in Chemical Kinetics," in I. Prigogine (ed.), *Advances in Chemical Physics*, Wiley-Interscience, New York, 1958.
8. K. B. Wiberg, *Chem. Rev.*, **55**, 713 (1955).
9. F. H. Westheimer, *Chem. Rev.*, **61**, 265–73 (1961).
10. W. H. Saunders, Jr., "Investigation of Rates and Mechanisms of Reactions," in S. L. Friess, E. S. Lewis, and A. Weissberger (eds.), *Technique of Organic Chemistry*, 2d ed. (revised), Wiley-Interscience, New York, 1961.

† A case where the secondary isotope effect more than compensates for the primary effect is the aqueous hydrolysis of borohydride ion. The species BH_4^- hydrolyzes more slowly in H_2O than does BD_4^- in H_2O.[25] In this case, the three nonreacting deuterium atoms cause a rate enhancement greater than the rate-decreasing effect of the deuterium in the bond that is breaking. The ratio $k(BH_4^-)/k(BD_4^-)$ has been found to be 0.70.

11. R. P. Bell, *The Proton in Chemistry*, Cornell University Press, Ithaca, N.Y., 1959, Chap. XI.
12. M. L. Bender, *Chem. Rev.*, **60**, 53–113 (1960).
13. W. P. Jencks and J. Carriuolo, *J. Am. Chem. Soc.*, **82**, 1778 (1960).
14. Cf. L. Larsson, *Svensk Kem. Tidskr.*, **70**, 405 (1958).
15. C. G. Swain and C. B. Scott, *J. Am. Chem. Soc.*, **75**, 141 (1953).
16. J. O. Edwards, *J. Am. Chem. Soc.*, **76**, 1540 (1954).
17. J. O. Edwards, *J. Am. Chem. Soc.*, **78**, 1819 (1956).
18. E. Grunwald and S. Winstein, *J. Am. Chem. Soc.*, **70**, 846 (1948).
19. J. E. Leffler, *J. Org. Chem.*, **20**, 1202 (1955); see also W. K. Wilmarth and N. Schwartz, *J. Am. Chem. Soc.*, **77**, 4543 (1955).
20. R. T. M. Fraser, in S. Kirschner (ed.), *Advances in the Chemistry of the Coordination Compounds*, Macmillan, New York, 1961, p. 287.
21. G. S. Hammond, *J. Am. Chem. Soc.*, **77**, 334 (1955).
22. H. R. Ellison et al., *J. Am. Chem. Soc.*, **84**, 1824 (1962).
23. J. O. Edwards and R. G. Pearson, *J. Am. Chem. Soc.*, **84**, 16 (1962).
24. S. I. Miller, *J. Am. Chem. Soc.*, **81**, 101 (1959).
25. R. E. Davis, E. Bromels, and C. L. Kibby, *J. Am. Chem. Soc.*, **84**, 885 (1962).
26. O. Reitz, *Z. Physik. Chem. (Leipzig)*, **A176**, 363 (1936).

4

Nucleophilic Displacements

The type of reaction in which one electron donor replaces another in a substrate is termed a nucleophilic displacement and is quite common, both in inorganic and in organic chemistry. The general reaction

$$N + S—X \rightarrow N—S + X \tag{4-1}$$

where N is a nucleophile, S—X is the substrate, and X is a leaving group, can take place by a variety of mechanisms. In this chapter we shall discuss certain of the mechanistic types, the factors that influence rates of nucleophilic displacement, and some examples.†

4-1 TYPES OF MECHANISM

The simplest type of substitution is one without an intermediate; the rate step is the bimolecular collision of the nucleophile with the substrate, and the mechanism is a displacement of one donor by another. The reaction-coordinate diagram has a single

---

†Since this chapter was finished, two reviews on nucleophilic displacement reactions have appeared. The references are (1) R. F. Hudson, *Chimia*, **16**, 173 (1962) and (2) J. F. Bunnett, *Ann. Rev. Phys. Chem.*, **14**, 271 (1963).

maximum near the center, since bond breaking and bond forma-tion take place simultaneously. This step is given the symbol S_N2 in the Hughes-Ingold classification[1] of organic mechanisms. Both experimentally and theoretically, the S_N2 displacement on saturated carbon compounds is well understood. The general properties of this transition state, as, e.g., the relative positions of atoms, have been evaluated from studies of molecules with asymmetric centers, with tracer isotopes, and with special stereochemical arrangements. In the activated complex the atoms N, S, and X form a linear configuration, which presumably permits minimum repulsion while the N—S bond is forming and the S—X bond is breaking.

The second mechanistic type, symbolized S_N1, is a two-step process

$$S—X \rightarrow S + X$$
$$N + S \rightarrow N—S \tag{4-2}$$

in which the first step is a bond-breaking step that forms an electron-deficient intermediate. Because the first step must be endothermic and the second exothermic, it is expected that the first step will always be rate-determining. The solvolysis of *tert*-butyl chloride mentioned in Chapter 3 proceeds by this mechanism.

Another type of two-step mechanism involves addition of the nucleophile to the substrate, followed by loss of the leaving group. The general reaction is represented

$$N + S—X \underset{k_2}{\overset{k_1}{\rightleftharpoons}} N—S—X$$
$$N—S—X \overset{k_3}{\rightarrow} N—S + X \tag{4-3}$$

This mechanism predicts several rate effects, depending on the conditions. If $k_3 \gg k_2$, the species N—S—X is an intermediate beyond the transition state; the rate is first-order each in substrate and nucleophile; and the rate constant reflects a large amount of bond formation (a strong dependence on the nature of the nucleo-phile) but little dependence on bond breaking (an insensitivity to the nature of the leaving group). If, on the other hand, $k_2 \gg k_3$, the species N—S—X is an intermediate prior to the transition state. More commonly, the condition $k_2 \simeq k_3$ obtains, and the reaction-coordinate diagram has two maxima plus a minimum near the center. If the energy minimum is shallow, the rate behavior is virtually indistinguishable from the typical S_N2 type of displacement; on the other hand, if the minimum is deep, the intermediate may be sufficiently stable to be identified.

Less common types of nucleophilic displacement mechanism involve nucleophilic catalysis of solvolysis and general-base catalysis of solvolysis. Both types of mechanism are important.

4-2 RATE SCALES

In terms of the transition-state theory, the rate constants reflect the equilibrium constant for the binding of nucleophile and substrate in the activated complex. Because a variety of factors govern the order of transition-state stabilities, several nucleophilic orders have been observed.

Much progress in sorting out factors and orders has been made empirically by the use of LFER. The Brönsted equation for general-base catalysis is a correlation of the rates of nucleophilic displacement on hydrogen with the basicities of the nucleophiles; these basicities are given in Table 4-1. Rates of nucleophilic displacement on some other substrates have also been found to follow these basicities. The correlation was most successful when a series of similar nucleophiles such as oxygen anions were employed; it was poorest and sometimes nonexistent when different types of nucleophilic atoms were employed.

Swain and Scott[2] set up a reactivity scale that correlates rates of displacement on the standard substrate, methyl bromide, by a variety of nucleophiles. Their relative values for nucleophilic reactivity, the n scale, are given in Table 4-1. On the other hand, Edwards[3, 4] suggested the correlation of rate data with independent (nonkinetic) properties of the nucleophile and pointed out that by using variable amounts of two independent properties of the nucleophile, it is possible to correlate the data for different substrates having different orders of nucleophilic character. His first equation[3] uses basicity and electrode potential as the two nucleophilic constants and therefore requires two substrate constants. Edwards' second equation[4] correlates nucleophilic character with basicity and polarizability and has the form

$$\log (k/k_0) = \alpha P + \beta H \tag{4-4}$$

where (k/k_0) is the rate relative to water, P is defined as log (R_N/R_{H_2O}) with R representing molar refractivity, and H is a function of basicity $(pK_a + 1.74)$. The coefficients α and β are

Table 4-1 *Orders of nucleophilic strength*

Nucleophile	$pK_a{}^a$	n^b	$E_n{}^c$	P^d
S⁻	12.9		3.08	0.611
SO₃⁻	9.1		2.57	
S₂O₃⁻	1.9	6.36	2.52	
SC(NH₂)₂	0.4	(4.1)	2.18	
I⁻	(−10)	5.04	2.06	0.718
CN⁻	9.1		2.02^d	0.373
SCN⁻	(−0.7)	4.77	1.83	
C₆H₅NH₂	4.5	4.49	1.78	
NO₂⁻	3.4		1.73	
OH⁻	15.7	4.20	1.65	0.143
N₃⁻	4.7	4.00	1.58	
Br⁻	(−7)	3.89	1.51	0.539
NH₃	9.5		1.36^d	0.184
Cl⁻	(−4)	3.04	1.24	0.389
C₅H₅N	5.3	(3.6)	1.20	
CH₃CO₂⁻	4.7	2.72	0.95	
SO₄⁻	2.0	(2.5)	0.59	
F⁻	3.2		−0.27^d	−0.150
H₂O	−1.7	0.00	0.00	0.000

[a] For conjugate acid of nucleophile; parentheses denote rough values.
[b] Nucleophile scale of Swain and Scott.[2]
[c] Electrode-potential scale.[3]
[d] Polarizability scale.[4]

experimentally determined for each substrate; with their values suitably chosen, the equation can be made to fit a large amount of rate data.

There are, therefore, no less than four different scales for the correlation of rates of nucleophilic displacement reactions. These scales, presented in Table 4-1, are useful, but because of the variety of factors that influence nucleophilic reactivity, they do not cover all cases.

4-3   RATE FACTORS

The binding interaction of a nucleophile and a substrate in an activated complex is determined by the same fundamental laws that govern stabilities of other chemical adducts. Unfortunately, the present state of chemical theory does not allow calculation of such binding energies. It is necessary to rely on experiment and intuition to discover the important factors governing these energies.

The correlation of *basicity* with nucleophilic character stems from the fact that substitution reactions are generalized acid-base reactions. In Eq. (4-1), N and X are bases and S is an acid. The basicity of N (and similarly of X) is measured by the acid-base reaction

$$N + H_3O^+ \rightarrow NH^+ + H_2O$$

and it is to be expected that some correlation between the rate of attachment and the equilibrium binding strength will obtain. When the substrate is a proton, N is clearly interacting with a positive center. In the case of other substrates, N is interacting primarily with the S atom, which may have a positive charge or may be assuming one through the loss of X in the transition state. Therefore, to the extent that the substrate atom S has a localized positive charge similar to that of a proton, the basicity of the nucleophile will be important and the rates will correlate with basicity.

Some particles, such as thiourea, triphenylphosphine, and the iodide ion, are more nucleophilic than their basicities would warrant. It has been fashionable to ascribe this excess reactivity to *polarizability*, although a causal relationship has not been rigorously proved. It has been pointed out that polarizability is important in the stabilization of an activated complex by van der Waals forces. Also, a polarizable nucleophile can approach a substrate closely and yet by its "softness" avoid the electron-electron repulsion that results from the Pauli exclusion principle. Recently the necessity of low-energy, unfilled orbitals in a polarizable nucleophile has been emphasized,[5] along with the ability of such orbitals to stabilize an activated complex.

Those nucleophiles that are strongly basic are often not very polarizable; conversely, a large polarizable nucleophile may not be very basic. Just as some cations (Be^{++}, Ti^{4+}) form coordination compounds with the halide ligands in the stability order $F^- > Cl^- > Br^- > I^-$ and other cations (Ag^+, Hg^{++}) in the opposite order, $I^- > Br^- > Cl^- > F^-$, so some substrates will react primarily with a basic nucleophile (e.g., F^- on tetrahedral phosphorus) and others with a polarizable nucleophile (e.g., I^- on peroxide oxygen).

Certain substrates will react with nucleophiles that are both basic and polarizable.

It has recently become apparent that alternation of the solvent[6,7] is another factor that affects the order of nucleophilic reactivities. The most basic ion is often the most strongly solvated in a protonic solvent. Therefore solvation tends to oppose the influence of basicity, for it is necessary to desolvate the nucleophile before it can react with the substrate. The energy of desolvation will, of course, contribute to the energy of activation, although this will in part be compensated for by a more positive entropy of activation. For example, hydroxylic solvents such as water and the alcohols bind to the halide ions through hydrogen bonding in the order $F^- > Cl^- > Br^- > I^-$. This strong solvation of the ground state is one reason why fluoride ion, the most basic halide ion, is a poor nucleophile whereas iodide ion is a good nucleophile in displacements on saturated carbon in aqueous solution. In solvents that cannot hydrogen-bond (e.g., acetone and dimethylformamide), the rate order is apparently reversed, for chloride ion reacts more rapidly than iodide.[6,7] Gould[8] had previously pointed out that the correlation of nucleophilic reactivities with electrode potentials is probably linked to the solvation energies of the anions.

It has been found that some nucleophiles show reactivities above those expected from their basicities and polarizabilities. These nucleophiles, among them peroxyanions and hydroxylamine, are characterized by the presence of one or more unshared electron pairs on an atom adjacent to the nucleophilic atom. It has been proposed[5] that the excess reactivity shown by this class of reagent be called the *alpha effect*, with reference to the pair of electrons on the alpha atom. Although there is no firm explanation for this effect, it probably stems from partial relaxation in the transition state[5] of an electronic repulsion present in the ground state of the nucleophile. The magnitude of the alpha effect is surprising; for example, perhydroxide ion OOH^- is several orders of magnitude more reactive than hydroxide ion, even though the hydroxide ion is far more basic. This case will be discussed in Chapter 5.

Some of the other factors that influence nucleophilic reactivity have been listed by Jencks and Carriuolo.[9] These include (a) hydrogen bonding; (b) proton transfer and general-acid catalysis; (c) electrostatic effects; (d) steric effects; (e) resonance; and (f) relative bond strengths to protons and to other substrate atoms. Although these factors are important in specific cases, they are of less general relevance than basicity, polarizability, and alpha and solvation effects. In summation, the factors that determine the reactivity of a nucleophile in attack on a certain substrate will include general factors, such as basicity, and particular factors, such as the possibility of pi bonding between the nucleophile and the substrate atom.

4-4 ORDERS OF NUCLEOPHILIC CHARACTER

In this section, experimental observations on rates of second-order nucleophilic substitution reactions involving a variety of nucleophiles and substrates will be presented. For comparison, organic substrates will be included. Most of these data have been obtained from the paper by Edwards and Pearson[5] in which further references may be found.

Rate data for nucleophilic displacement on hydrogen are obtained from general-base-catalyzed reactions; data from general-acid catalysis can also be used since microscopic reversibility presumably holds. As expected, the rate constants follow the equilibrium basicity scale rather well, and good Brönsted plots are obtained. However, Bell[10-12] has pointed out some significant exceptions. For example, hydroxide ion is kinetically less reactive than one would expect from its basicity,[10] whereas oximate ions, which have a free electron pair on the alpha atom, react abnormally rapidly.[22] Strong electron delocalization in the anion has an adverse rate effect, as may be noted with the anions of pseudoacids like nitromethane.[11]

In the remarkable paper by Jencks and Carriuolo,[9] a variety of nucleophiles were reacted with p-nitrophenyl acetate. In these cases, which represent nucleophilic attack at carbonyl carbon, the order of nucleophilic strength correlates with basicity, although large deviations were noted for some nucleophiles, including those with a free electron pair on the alpha atom. A tetrahedral intermediate is apparently involved in such cases[13] as in most cases of nucleophilic displacement on unsaturated carbon.

Attack at tetrahedral phosphorus will be discussed in Section 4-5; it is sufficient to say here that the order of nucleophilic strength is similar but not identical to that observed with p-nitrophenyl acetate.

Nucleophilic attack at boron occurs both when the substrate atom is trigonal and when it is tetrahedral. With trigonal R_2BX compounds, the rate order is $OH^- > OR^- > NH_3 > R_2NH \simeq RS^-$, and basicity is an important factor. The order of attack on the tetrahedral boron compound H_3NBF_3 appears to be $OH^- > F^- > H_2O$, with Cl^- showing no effect. Again basicity is a prime factor.

In attack on sulfur atoms, there is little data from which a quantitative scale of nucleophilic character can be assessed; but assuming that nucleophilic power correlates roughly with binding strength for any one substrate atom, the following orders are deduced from data on competitive reactions and equilibrium data. For tetrahedral sulfur (as in p-nitrophenyl p-toluenesulfonate and in neopentyl p-toluenesulfonate), the rough order is $RS^- > R_3P > C_6H_5S^- \simeq CN^- > SO_3^= > OH^- > S_2O_3^= > SC(NH_2)_2 > SCN^- > Br^- > Cl^-$. Apparently, nucleophilic attack on sulfur requires both polarizability and basicity components; it is interesting to note that basicity is more important in attack on tetravalent sulfur than on divalent sulfur.

The order of attack on aromatic carbon (as, for example, on 2,4-dinitrochlorobenzene) is $C_6H_5S^- \simeq CH_3O^- > C_5H_{10}NH > C_6H_5O^- > N_2H_4 > OH^- > C_6H_5NH_2 > Cl^- > CH_3OH$. The order $C_6H_5NH_2 > NH_3 > I^- > Br^-$ is also reported. Apparently, both basicity and polarizability are required here. It is worth noting that the discrimination between nucleophiles is large; for example, methoxide ion is about 10^{13} times more reactive than the methanol molecule itself.

The case of saturated carbon is, of course, the one most carefully studied; it was mentioned before and will be discussed further in Chapter 5. The order in this case can be considered to result from a combination of polarizability and basicity factors, and closely follows that of electrode potentials for oxidation of the nucleophiles.[3]

Compounds of the type NH_2X will react with nucleophiles by displacement at nitrogen when X is a good leaving group (weakly basic anion such as Cl^- or SO_4^-). Although there is little quantitative data available, polarizability is obviously important and basicity somewhat less so.

The data on nucleophilic attack on peroxide oxygen are summarized in Chapter 5. It is apparent that the observed scale depends largely on polarizability, and the discrimination between nucleophiles is large. A similar scale has been observed for nucleophilic attack on Pt(II) compounds, but it reflects only a small discrimination among nucleophiles. Attack on these square-planar complexes will be discussed in Section 4-6.

The above listings cover only a portion of the known displacement reactions in which a nucleophile attacks an inorganic center. Attacks on fluorine (in perchloryl fluoride), on chlorine (in hypochlorites), on bromine (in dehalogenation reactions), on iodine (in iodate ion), on silicon (in siliconium ions), and on transition metals (in octahedral complexes) have been reported. It is apparent that this type of reaction mechanism is common to atoms in all parts of the periodic table.

4-5   DISPLACEMENTS ON PHOSPHORUS

In this section nucleophilic displacements on tetrahedral phosphorus substrates will be covered in some detail. These compounds are important since many of them show physiological activity and have, therefore, been used as insecticides. Their chemistry, metabolism, and biological effects have been reviewed recently.[14, 15] Their *in vivo* activity has been traced to the inactivation of the cholinesterase enzyme. The four-coordinate phosphorus compounds react with the enzyme's active site, which is a nucleophilic atom or group capable of rapidly hydrolyzing the ester acetylcholine. The acetyl group is transferred to the active site and then, in turn, is rapidly hydrolyzed off to reactivate the enzyme. In the case of poisoning by a phosphorus compound, the active site is phosphorylated and the enzyme cannot be reactivated in the usual fashion.

Since the quantitative data on displacements in neutral four-coordinate phosphorus compounds are scattered in the literature, it is difficult to prepare a list of nucleophiles in order of their relative strengths. Table 4-2 presents some data along with comparative data for p-nitrophenyl acetate displacements. In the case of both substrates, there is a correlation between rate and nucleophile basicity, although there are some marked deviations as well.[16–18] In water, the order appears to be $OOH^- > OH^- \simeq OCl^- > NH_2OH > NO_2^- > N_3^- > H_2O$. In ethanol, the order $F^- > C_2H_5O^- > C_6H_5O^-$ has been observed. Sulfur nucleophiles such as $S_2O_3^=$ and $C_6H_5S^-$ do not seem to be particularly reactive, and the ions Cl^-, Br^-, and I^- are nonreactive. The reactivity of the fluoride ion is surprisingly high, as are the reactivities of oxygen nucleophiles with unshared electrons on the alpha atom.[17] Pi bonding between the nonbonded electron pairs (other than the pair forming the sigma bond) of oxygen or fluorine and the unfilled

Table 4-2 *Rates of nucleophilic replacements*

| Nucleophile | pK_a | Rate constants, liter mole^{-1} min^{-1} | |
		Carbonyl carbon[a]	Tetrahedral phosphorus[b]
HOO^-	11.5	2×10^5	1.0×10^5
Acetoximate	12.4	3.6×10^3	
Salicylaldoximate	9.2	3.2×10^3	1.5×10^3
OH^-	15.7	9×10^2	1.6×10^3
$C_6H_5O^-$	10.0	1×10^2	34
NH_2OH	6	1×10^2	1.3
OCl^-	7.2	1.6×10^3	7×10^2
$CO_3^=$	10.4	1.0	75
NH_3	9.2	16	
CN^-	10.4	11	
$C_6H_5S^-$	6.4		7.4×10^{-3}
$C_6H_5NH_2$	4.6	1.5×10^{-2}	
C_5H_5N	5.4	0.10	
NO_2^-	3.4	1.3×10^{-3}	
$CH_3CO_2^-$	4.8	5×10^{-4}	
F^-	3.1	1×10^{-3}	Very reactive[c]
$S_2O_3^=$	1.9	1×10^{-3}	Nonreactive
H_2O	-1.7	6×10^{-7}	1×10^{-6}

[a] *p*-Nitrophenyl acetate as substrate.[7]
[b] Methylisopropoxyphosphoryl fluoride (Sarin) as substrate.[16-18]
[c] Estimated from results obtained with similar substrates.

d orbitals on phosphorus may cause a low activation energy and the resultant high rate.

The role of basicity in nucleophilic strength varies considerably with the individual phosphorus compound attacked. With hydroxamate ions of varying basicity as nucleophiles, it has been found that the resultant Brönsted plots for displacement on Sarin (methylisopropoxyphosphoryl fluoride), TEPP (tetraethyl-pyrophosphate), and Tabun (dimethylaminoethoxyphosphoryl cyanide) have slopes of 0.9, 0.7, and 0.5, respectively. Further data reflecting this difference in response to nucleophile basicity is shown in Table 4-3, which lists rates for a variety of nucleophiles in reaction with two of the above phosphorus substrates. The

Table 4-3 *Influence of basicity on the reactivity of nucleophiles with Sarin and TEPP[a]*

Reagent	pK_a[b]	k_{TEPP}, liter mole^{-1} min^{-1}	k_{Sarin}, liter mole^{-1} min^{-1}	k_{TEPP}/k_{Sarin}
H_2O	-1.7	0.0017	0.0001	17
NH_2OH	6[c]	26	2.6	10
ClO^-	7.4	267	600	0.45
$C_6H_5CONHO^-$	8.8	160	1020	0.16
$CH_3COC(CH_3)NO^-$	9.3	16	380	0.043
HO_2^-	11.8	2180	94,000	0.023
HO^-	15.7	21	2,000	0.011

[a] Data from Ref. 17.

[b] For ionization of conjugate acid of listed reagent.

[c] Refers to nitrogen basicity, whereas the oxygen function is the reactive site in attack on phosphorus.

ratios between rate constants (last column) indicate that the two phosphorus compounds show differences in their responses to basicity not accounted for by their intrinsic reactivities.

The data in Table 4-4 clarify the reason for this difference in response to basicity. The three phosphorus substrates are of the type (I), where R_1 and R_2 can be ethyl and/or ethoxy. When the

$$R_1\diagdown \underset{R_2\diagup}{P}\diagup^{\diagup O}_{\diagdown F}$$

(I)

Table 4-4 *Effect of conjugative substituents on the importance of basicity in phosphoryl substitution[a]*

Compound	$k_{HO^-}/k_{CH_3COCH:NO^-}$
$(C_2H_5)_2P(O)F$	95
$(C_2H_5O)(C_2H_5)P(O)F$	7.5
$(C_2H_5O)_2P(O)F$	1.8

[a] Data from Ref. 17. See data in Table 4-3 for pK_a values and for similar rate ratios.

groups are ethoxy, the empty d orbitals on phosphorus are partially occupied as a result of pi bonding with the filled p orbitals on oxygen. There can be less bonding between the incoming oxygen nucleophile and the phosphorus in the transition state; nucleophile basicity is therefore of less importance in the transition state. As might be expected, the replacement of an alkyl group on phosphorus by the homomorphous alkoxy group usually lowers the reactivity, although as the data in Table 4-3 shows this is not always true.

Since both H_2O and OH^- are nucleophiles in displacement on tetrahedral phosphorus, the plot of the log of the rate constant (pseudo-first-order) against pH should show a horizontal region at low pH and a positive slope at high pH; such is certainly the case for many phosphorus compounds. However, when the leaving group is strongly basic, as are fluoride ion or dimethylamine, acid catalysis is also observed. Acid catalysis of solvolysis has been seen in the cases of Tabun, Sarin, and diisopropyl fluorophosphate (DFP). The pH profile of the latter is presented in Figure 4-1; segments labeled A, B, and C represent acid-catalyzed solvolysis, neutral solvolysis, and base-catalyzed solvolysis, respectively.

The leaving group has a significant effect on the rate of hydrolysis. For compounds of the type (II), where X is a leaving anion,

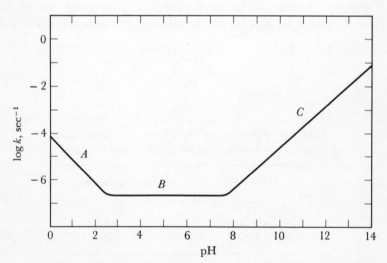

Figure 4-1   *The rate of hydrolysis of diisopropyl fluorophosphate as a function of pH at 25°C.*

$$
\begin{array}{c}
\text{CH}_3 \quad\ \text{O} \\
| \qquad\ || \\
\text{H—C—O—P—X} \\
| \qquad\ | \\
\text{CH}_3 \quad \text{CH}_3
\end{array}
$$

(II)

the rates follow the order $p\text{-NO}_2\text{—C}_6\text{H}_4\text{O}^- > p\text{-Cl—C}_6\text{H}_4\text{O}^- >$ $\text{C}_6\text{H}_5\text{O}^- \simeq p\text{-CH}_3\text{O—C}_6\text{H}_4\text{O}^-$.[18] The orders $\text{Cl}^- \gg \text{F}^-$ and $\text{RS}^- > \text{RO}^-$ have also been reported.[16] Since the leaving group should depart more readily as it becomes less basic, these orders are expected in cases where some bond breaking occurs in the transition state.

The detailed mechanism by which the phosphorus substrate is solvolyzed is more complicated than it would appear to be at first sight. Some of the many unanswered questions in this area will be discussed in Chapter 10 of this book. Suffice it to say here that for the majority of these substrates, the distinction between specific-base catalysis, general-base catalysis, and nucleophilic catalysis has not been made; possibly all three mechanisms are important in some cases.

4-6 DISPLACEMENTS IN PLANAR COMPLEXES

As will be seen in Chapter 6, one of the more vexing mechanistic problems involves ligand replacement in octahedral complexes. No such problem occurs in the square-planar complexes, where excellent evidence for a simple displacement mechanism[19] exists.

Starting with a common substrate such as the tetrachloro-platinate(II) ion, it is found that a variety of ligands will replace one or more chloride ions. The rate is first-order each in complex and ligand, and the trend of nucleophilic strength is found to be $\text{R}_3\text{P} \simeq (\text{NH}_2)_2\text{CS} \simeq \text{SCN}^- \simeq \text{I}^- > \text{N}_3^- > \text{NO}_2^- > \text{C}_5\text{H}_5\text{N} >$ $\text{C}_6\text{H}_5\text{NH}_2 > \text{olefin} \simeq \text{NH}_3 \simeq \text{Br}^- > \text{Cl}^- > \text{OH}^- \simeq \text{H}_2\text{O} \simeq \text{F}^-$. This trend is similar to that observed with peroxide oxygen (see Chapter 5), for in both cases olefins and phosphines are good nucleophiles whereas hydroxide and ethoxide ions are poor nucleophiles. For both substrates, polarizability is a dominant factor in nucleophilic strength; basicity plays no important role.

The nature of the platinum complex influences the quantitative order. For PtCl_4^- and $\text{PtNH}_3\text{Cl}_3^-$, nitrite ion is more than a thousand times as reactive as chloride ion; for $[\text{Pt(dien)Cl}]^+$, where dien is diethylenetriamine, the reactivity ratio is only 5.

From these and other observations, it appears that as the negative charge on the complex increases, polarizability enhances nucleophilic character toward Pt(II).

Because there are five available orbitals (one d, one s, and three p) on Ni(II), Pd(II), Pt(II), and Au(III) and only four are used by the ligands, the possibility of a five-coordinate intermediate must be considered. There is, however, no strong evidence for such an intermediate.

At least in the case of Pt(II), discrimination between nucleophiles is small. For example, pyridine barely competes with solvent water as a nucleophile, although subsequently the water molecule is quickly replaced by a pyridine molecule.[19, 20] Presumably the mechanism

$$PtCl_4^- + H_2O \rightarrow PtCl_3OH_2^- + Cl^- \qquad \text{slow}$$

$$PtCl_3OH_2^- + py \rightarrow PtCl_3py^- + H_2O \qquad \text{fast}$$

competes with the direct displacement of chloride ion by pyridine. Experimental evidence for these competing paths exists: the first-order loss of $PtCl_4^-$ is linearly dependent on the concentration of pyridine, yet there is a definite rate of reaction even when the pyridine concentration is extrapolated to zero. The rate law is

$$v = k_1[PtCl_4^-] + k_2[PtCl_4^-][\text{pyridine}] \qquad (4\text{-}5)$$

The rate of reaction is sensitive to the nature of the complex to an extent beyond the effect of the leaving group. The *trans* complexes $[M(Cl)(o\text{-tolyl})(PEt_3)_2]$ react with pyridine as the entering nucleophile and chloride ion as the leaving group. The relative rates of equilibrium attainment are approximately 5,000,000: 100,000:1 where M is Ni(II), Pd(II), and Pt(II), respectively.[20] The marked difference between the rates of the Ni and Pt complexes is consistent with the known relative ease of expansion for nickel complexes to higher coordination numbers (usually six).

Strong directing groups already present can guide the course of substitution in square-planar complexes. The ability of a group to direct substitution into the position *trans* (opposite) to itself has been called the *trans effect*, and such a group has a marked influence on the rate of reaction.[21] This can be illustrated by the reactions

$$\begin{array}{cc} \text{Cl} & \text{C}_2\text{H}_4 \\ | & | \\ \text{Cl}-\text{Pt}-\text{NH}_3^- + \text{C}_2\text{H}_4 \rightarrow \text{Cl}-\text{Pt}-\text{NH}_3 + \text{Cl}^- \\ | & | \\ \text{Cl} & \text{Cl} \end{array}$$

$$
\begin{array}{c}
\text{Cl} \\
| \\
\text{Cl—Pt—C}_2\text{H}_4^- + \text{NH}_3 \\
| \\
\text{Cl}
\end{array}
\rightarrow
\begin{array}{c}
\text{Cl} \\
| \\
\text{H}_3\text{N—Pt—C}_2\text{H}_4 + \text{Cl}^- \\
| \\
\text{Cl}
\end{array}
$$

In the first case, chloride *trans* to chloride is replaced; in the second, the chloride *trans* to ethylene is replaced. This and other data show that ethylene has a greater *trans* effect than chloride and chloride a greater effect than ammonia. The directing nature of the *trans* effect has been used in the synthesis of complexes of Pt(II); for example, the three geometric isomers of $[\text{Pt}(\text{Cl})(\text{Br})(\text{NH}_3)(\text{py})]$ have been made successfully by employment of successive ordered substitutions.

Extensive qualitative studies of the above type have made it possible to list the ligands in order of their ability to direct an incoming nucleophile into the *trans* position. For increasing *trans* effect, the rough order[21] is $\text{H}_2\text{O} < \text{OH}^- < \text{NH}_3 < \text{RNH}_2 <$ pyridine $< \text{Cl}^- < \text{Br}^- < \text{SCN}^- \simeq \text{I}^- \simeq \text{NO}_2^- \simeq \text{SO}_3\text{H}^- \simeq \text{PR}_3 \simeq \text{R}_2\text{S} \simeq \text{SC}(\text{NH}_2)_2 < \text{NO} \simeq \text{CO} \simeq \text{C}_2\text{H}_4 \simeq \text{CN}^-$. More recently the added order $\text{Cl}^- < \text{C}_6\text{H}_5^- < \text{CH}_3^- < \text{H}^- \simeq \text{PR}_3$ was observed.[20] The groups at the CN^- end of the series are said to have a high *trans* effect and those at the water end a low *trans* effect. It is interesting to note that the ligands at the high end of the series are those that can form pi bonds by accepting *d*-orbital electrons from platinum; however, this cannot be the complete story since the methide and hydride ions also have strong effects but cannot form pi bonds.

As may be seen in Table 4-5, the *trans* effect is not primarily a thermodynamic effect. Large variations in rate but only small

Table 4-5 *Comparison of rate constants with concentration equilibrium constants for reactions between pyridine and some platinum(II) complexes*[a]

Complex	k_1, min^{-1}	K_{eq}
trans-[Pt(H)(Cl)(PEt$_3$)$_2$]	1.1 (at 0°C)	0.03
trans-[Pt(Cl)(Me)(PEt$_3$)$_2$]	1×10^{-2}	0.15
trans-[Pt(Cl)(Ph)(PEt$_3$)$_2$]	2×10^{-3}	0.02
cis-[Pt(Cl)(Me)(PEt$_3$)$_2$]	3.6 (at 0°C)	0.3
cis-[Pt(Cl)(Ph)(PEt$_3$)$_2$]	2.3	0.15

[a]Data (in ethanol solvent at 25°C) from Ref. 20.

differences in equilibrium constant are observed.[20] It can therefore be concluded that the *trans* effect is exerted in the transition state and that any explanation for it must be kinetic in nature. Two explanations that, taken together, fit the data are the polarization theory of Grinberg[22] and the pi-bonding theory of Chatt[23] and Orgel.[24]

Grinberg[22] pointed out that the *trans*-effect order for the halides is $I^- > Br^- > Cl^-$ and that the weakening of the metal-ligand bond opposite the *trans*-directing group could be attributed to the change in polarizability of the group. As may be seen in Figure 4-2(*a*), the dipoles induced by the central metal ion in the ligand will cancel each other and the resultant dipole will be zero if all the ligands are the same. If, however, one of the ligands is different from the other three [as in Figure 4-2(*b*)], there will be a resultant dipole in the central ion. If Y is more polarizable than X, then the central ion in turn becomes polarized in such a way that it is less positive at a point directly opposite to Y than it is at a point *cis* to Y. Therefore the M—X bond *trans* to Y will be weaker than M—X bonds *cis* to Y. Since initial and final states have equal probability of *trans* repulsion by polarization, the equilibrium constants are not affected, but the rate constants are indeed affected by a polarizable ligand. It is worth noting that the order of *trans* effect for simple ligands nicely parallels their order of nucleophilic strength; this is to be expected since polarizability should influence both.

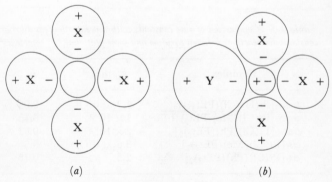

(*a*)   (*b*)

Figure 4-2   *Grinberg's polarization theory of the **trans** effect.*

The idea of pi bonding in metal coordination compounds was proposed by Pauling[25] to account for the high stability of cyanide and carbon monoxide complexes. It was later suggested[23, 24] that there was a correlation between the tendency of a ligand to remove d-orbital electrons from the metal by pi bonding and the magnitude of the *trans* effect. Orgel, in his theory,[24] pointed out that the transition state for a nucleophilic displacement on a square-planar complex has a distorted trigonal-bipyramid structure (III), which

$$L = Pt \overset{\displaystyle A}{\underset{\displaystyle A}{\vert}} \overset{\nearrow Y}{\searrow_X}$$

(III)

can be stabilized by double bonding to the *trans* ligand L. The two A groups, which are *cis* to L both in the initial and in the final states, form the apexes of the trigonal-bipyramid transition state. This transition state results from the approach of Y on one side of the plane and a compensating motion of X into the opposite side. The ligand L enhances the stability of such a trigonal bipyramid by withdrawal of nonbonding electrons in the d_{xz} orbital (assuming the square complex to be in the xy plane) away from the incoming nucleophile. A similar but more detailed explanation was proposed by Chatt.[23]

Surprisingly, the more stable the complex $PtX_4^=$, the more rapid is its exchange with unbound ligand X^-. Some data in support of this are presented in Table 4-6. Since the reaction rates

Table 4-6   *Rates of ligand exchange between $PtX_4^=$ and X^-*

X^-	Stability constant	$t_{1/2}$, min[a]
CN^-	10^{35}	1
I^-	10^{25}	5
Br^-	10^{21}	8
Cl^-	10^{16}	280

[a]Concentrations about 0.1 M; data from A. A. Grinberg and E. L. Nikol'skoya, *Zh. Prikl. Khim.*, **22**, 542 (1949); **24**, 893 (1951).

Table 4-7 *Rates for some* $[Pt(dien)X]^+$ *and* $[Pd(dien)X]^+$
complexes with pyridine at 25°C[a]

	k_{obs}, min^{-1}	
X$^-$	$[Pt(dien)X]^+$	$[Pd(dien)X]^+$
NO$_3^-$	fast[b]	
Cl$^-$	2.1×10^{-3}	fast[b]
Br$^-$	1.4×10^{-3}	fast[b]
I$^-$	6.0×10^{-4}	2.0^c
N$_3^-$	5.0×10^{-4}	
SCN$^-$	1.8×10^{-5}	2.5
NO$_2^-$	3.0×10^{-6}	2.0
CN$^-$	1.0×10^{-6}	

[a]Data from F. Basolo, H. B. Gray, and R. G. Pearson, *J. Am. Chem. Soc.*, **82**, 4200 (1960); dien is diethylenetriamine.
[b]Too fast to measure.
[c]At 0°C.

are first-order in complex but zero-order in entering ligand, the solvent probably is acting as nucleophile and is constant for all cases. Also, the rapid rates for the more stable complexes suggest that a five-coordinate species (kinetic intermediate and/or transition state) is stabilized by the *trans* effect, which follows the same order as bond strength in this series.

In a series where the *trans* effect is constant but leaving-group ability can vary, there is certainly evidence that the more stable the ligand-metal bond, the slower is the rate. Some rates that illustrate this point are listed in Table 4-7.

4-7 DISCUSSION

A variety of factors influencing nucleophile reactivity with substrates have been mentioned, and it appears possible to sort out three of the four main sources of high nucleophilicity. Two of the four, polarizability and desolvation phenomena, behave in a similar fashion and have been separated in only a few cases; they will be combined here, therefore, under polarizability. The sources of the alpha effect, basicity and polarizability, are apparently independent of each other. The alpha effect is observed mainly where bond making is significant; that is, it is particularly common with those substrates that strongly discriminate between nucleophiles. However, further experimental work will be necessary before the range and true nature of the alpha effect are known.

The examples given previously show that the various substrates differ in their susceptibility to basicity and polarizability. Substrates containing carbonyl carbon, tetrahedral phosphorus, trigonal boron, and the proton seem to depend almost entirely on basicity, whereas substrates containing oxygen, fluorine, and platinum seem to depend almost entirely on polarizability. Aliphatic tetrahedral carbon, aromatic carbon, and trivalent nitrogen depend on both factors, with polarizability rather dominant.

Theoretical discussion dealing with factors that determine nucleophilic reactivity[5] permits a satisfactory understanding of these variations. To the extent that the interaction between a substrate and a nucleophile in the transition state resembles the interaction of a proton with the nucleophile, we should find nucleophile basicity important. The characteristics of the proton are a high positive charge and an absence of outer electrons. Therefore, basicity will predominate with the substrates that most nearly fulfill these requirements. In ester hydrolysis, the structure available for the transition state is $=C^+—O^-$. It presents a more positive carbon to the incoming nucleophile and, equally important, one less pair of electrons attached to carbon as compared with a substrate such as an alkyl halide. In the compounds of tetrahedral phosphorus, there is both a high positive charge and a set of empty d orbitals on the central atom. As pi bonding from oxygen to the central atom reduces the positive charge and fills up these orbitals, susceptibility to basicity decreases. Trivalent boron is positive in nature and has an empty p orbital as well. The empty orbitals in all these cases have a twofold function: they reduce the number of repelling electron pairs on the substrate atom, and they provide a positive site for the acceptance of electrons from the nucleophile.

The cases in which polarizability of the nucleophile is the chief factor show a characteristic pattern. The central element in the substrate is electronegative, often negatively charged in the ground state, and has a number of outer-orbital electrons. The first two factors cause basicity to play a minor role, and the last factor ensures that polarizability will be important. Recall that an important property of a highly polarizable nucleophile is the ability to provide a low-energy, empty orbital to accommodate electrons from the substrate. In the cases of oxygen and fluorine, the central substrate atom has a full set of p electrons. In the case of platinum(II), there is a full set of d electrons except for one vacant d orbital in the plane of the complex; the eight d electrons project in all directions from which a nucleophile could reasonably approach the central atom.

The dual function of a polarizable nucleophile, donation of electrons to the substrate atom in a sigma bond and acceptance of electrons from the substrate atom in a pi bond, has led to the suggestion that such nucleophiles be called *biphilic reagents*.[26] Existing evidence indicates that biphilicity is important for complexes of the transition metal ions, which have a relatively low positive charge and many d electrons. In the metal carbonyls, for example, only a biphilic reagent such as carbon monoxide itself or trialkylphosphine can easily displace an attached carbon monoxide molecule. This may be due to thermodynamic factors, but data in a few cases show that simple nucleophilic displacements do occur.[27] Hydroxide ion is often surprisingly poor as a reagent. For example, in the case of *cis*- or *trans*-$[Rh(en)_2Cl_2]^+$, the rate of reaction with hydroxide ion is hardly greater than the rate of reaction with water. This is unexpected in view of the high positive charge of rhodium(III). The data for platinum(II) cited earlier also show hydroxide ion to be a poor reagent. For metal atoms with no outer d electrons and with a high charge, hydroxide ion again becomes a good reagent. This is true for chromium(VI), silicon(IV), germanium(IV), and tin(IV).

A biphilic reagent is not the same thing as an *ambident reagent*.[28] The latter has two different nucleophilic sites, as, for example, the oxygen and nitrogen atoms in the nitrite ion. In any given transition state, only one site is involved. In most ambident reagents, one nucleophilic site is more basic and the other is more polarizable or presents an electronic structure with an empty orbital. If the substrate S—X resembles a proton, the more basic site will react; this would be the case if an intermediate similar to a carbonium ion were formed. The more polarizable site will react with a substrate that favors polarizability over basicity, as does a primary alkyl halide.

Displacement reactions on alkyl halides and similar compounds require both basicity and polarizability, with the latter more important. This results naturally from the structure of the transition state. The central carbon atom is somewhat positive, but not greatly so. The orbitals surrounding carbon are filled, but the electrons are somewhat removed from the critical region by bonding to other atoms or groups. The general situation is clearly intermediate between that of a proton as the substrate and that of a peroxide oxygen atom as a substrate, but rather closer to the latter case. As one goes from tetrahedral carbon compounds, R_3CX, across the periodic table to R_2NX, ROX, and finally FX, one expects and finds that polarizability becomes more important and basicity less important. The fact that tetrahedral boron compounds, R_3BX, depend more on basicity than does R_3CX is also expected.

It seems reasonable to conclude that much of the data related to nucleophilic displacements on inorganic substrates can be understood in terms of a few important factors that influence the stabilities of the transition states. One must not, however, lose sight of the individual nature of each case. The fact that olefins are good nucleophiles to Pt(II) compounds and also to peroxide oxygen compounds is undoubtedly linked to the stable ring structure found in both products. The fact that fluoride ion is an exceptional nucleophile to tetrahedral phosphorus substrates is linked to the strong bond formed between phosphorus and fluorine. Loosely speaking, the transition state for a nucleophilic displacement reaction sees the product toward which it is proceeding.

References

1. C. K. Ingold, *Structure and Mechanism in Organic Chemistry*, Cornell University Press, Ithaca, N.Y., 1953, p. 306ff.
2. C. G. Swain and C. B. Scott, *J. Am. Chem. Soc.*, **75**, 141 (1953).
3. J. O. Edwards, *J. Am. Chem. Soc.*, **76**, 1540 (1954).
4. J. O. Edwards, *J. Am. Chem. Soc.*, **78**, 1819 (1959).
5. J. O. Edwards and R. G. Pearson, *J. Am. Chem. Soc.*, **84**, 16 (1962).
6. S. Winstein et al., *Tetrahedron Letters*, **1960** (No. 9), 24.
7. A. J. Parker, *J. Chem. Soc.*, **1961**, 1328.
8. E. S. Gould, *Mechanism and Structure in Organic Chemistry*, Holt, Rinehart, and Winston, New York, 1959, p. 260.
9. W. P. Jencks and J. Carriuolo, *J. Am. Chem. Soc.*, **82**, 1778 (1960).
10. R. P. Bell, *Acid-Base Catalysis*, Clarendon Press, Oxford, 1941, p. 92.
11. R. P. Bell, *J. Phys. Chem.*, **55**, 885 (1951).

12. R. P. Bell, *The Proton in Chemistry*, Cornell University Press, Ithaca, N.Y., 1959.
13. M. L. Bender, *Chem. Rev.*, **60**, 53 (1960).
14. R. D. O'Brien, *Toxic Phosphorus Esters*, Academic Press, New York, 1960.
15. *Chem. Soc. (London), Spec. Publ.*, **1957** (No. 8).
16. Cf. L. Larsson, *Svensk Kem. Tidskr.*, **70**, 405 (1959).
17. A. L. Green, G. L. Sainsbury, B. Saville, and M. Stansfield, *J. Chem. Soc.*, **1958**, 1583.
18. J. Epstein, private communication, 1961.
19. F. Basolo and R. G. Pearson, *Mechanism of Inorganic Reactions*, Wiley, New York, 1958, p. 172ff.
20. F. Basolo, J. Chatt, H. B. Gray, R. G. Pearson, and B. L. Shaw, *J. Chem. Soc.*, **1961**, 2207.
21. The *trans* effect is reviewed in Ref. 19, pp. 172–192.
22. A. A. Grinberg, *Ann. inst. platine (U.S.S.R.)*, **10**, 58 (1932).
23. J. Chatt, L. A. Duncanson, and L. M. Venanzi, *J. Chem. Soc.*, **1955**, 4456; see also previous papers referenced therein.
24. L. E. Orgel, *J. Inorg. Nucl. Chem.*, **2**, 137 (1956).
25. L. Pauling, *The Nature of the Chemical Bond*, Cornell University Press, Ithaca, N.Y., 1940, p. 252.
26. R. G. Pearson, H. B. Gray, and F. Basolo, *J. Am. Chem. Soc.*, **82**, 787 (1960).
27. F. Basolo and A. Wojcicki, *J. Am. Chem. Soc.*, **83**, 520, 525 (1961).
28. N. Kornblum, R. A. Smiley, R. K. Blackwood, and D. C. Iffland, *J. Am. Chem. Soc.*, **77**, 6269 (1955).

5

Nonradical Mechanisms for Peroxide Reactions

The low dissociation energy (about 35 kcal mole^{-1}) of the oxygen-oxygen single bond in peroxides allows this bond to be severed homolytically into free radicals with relative ease. This low energy and the strong oxidizing power of peroxides allow the intrusion of trace metals, particularly transition metals, to act as catalysts for reactions. Such complications as free radical formation and catalysis by adventitious metal ions are more or less serious in all peroxide reactions.

It can be difficult to determine whether a particular mechanism involves free radicals. The evidence for and against free radicals in mechanisms has been obtained by a variety of techniques. When free radicals from peroxides are present, the addition of a vinyl monomer will usually change the rate of reaction as well as the kinetic law. If a reducing radical is formed, the rate may depend on the amount of oxygen gas present, since oxygen combines with certain radicals at a very high rate. If the enthalpy of activation is less than one-half the dissociation energy of the weakest bond (usually the peroxide bond), the energetics are inconsistent with a free radical mechanism, as will be seen in Chapter 9. The rate laws for free radical mechanisms often have half-integral orders, whereas those for polar reactions normally have integral orders.

Trace-metal catalysis can be a serious problem in peroxide mechanism studies. Amounts as small as 10^{-9} M in certain transition metal cations are known to influence the rates of some peroxide reactions. Since the solvents, the reactants, and even the peroxides themselves contain small amounts of impurities, elimination of these impurities is a major problem. Addition of powerful complexing agents such as ethylenediaminetetraacetic acid, which tie up the catalyzing metal ion, can render these ions ineffective in some cases. The best evidence that trace metals are involved in a reaction mechanism is an unusual rate law (for example, first-order in peroxide but zero-order in reducing agent) coupled with erratic rate constants.

The mechanisms to be discussed here have kinetic patterns that are not consistent with the behavior of free radicals or with catalysis by trace metals. It is felt, therefore, that these reactions go by *polar* mechanisms; that is, the mechanisms involve simple atom- or ion-transfer steps. This chapter will cover several peroxide and hypohalite reactions that have unusual and interesting polar mechanisms.

5-1 NUCLEOPHILIC DISPLACEMENTS

In the previous chapter, it was seen that one pathway for substitution in inorganic compounds involves a nucleophilic displacement step. That such an S_N2 step occurs in some oxidations by peroxides[1-3] is not surprising, for reducing agents are species with available electrons just as are nucleophiles. The rate law for many of these oxidations is

$$v = k[ROOR'][N] \tag{5-1}$$

where ROOR′ is a peroxide and N is a nucleophile. As may be seen in Table 5-1, the rate constants are often conveniently measurable, the enthalpies of activation are low, and the entropies of activation are negative.

Some of the reducing agents that are oxidized by peroxides through such a displacement mechanism are chloride, bromide,

Table 5-1 *Some representative reactions of peroxides with nucleophiles*

R'OOR	N	Solvent	k, liter mole^{-1} sec^{-1}, 25°C	$\Delta H^{\ddagger}$	$\Delta S^{\ddagger}$	Note
H_2O_2	I^-	H_2O	0.69	12.8	-16	a
H_2O_2	SCN^-	H_2O	5.2×10^{-4}	14.9	-25	b
H_2O_2	$S(CH_2C_8H_4Cl)_2$	$2\text{-}C_3H_7OH$	1.91×10^{-4}	16.6	-24	c
$(CH_3)_3COOH$	$S(CH_2CH_2OH)_2$	H_2O	1.4×10^{-4}	14.2	-29	d
$(CH_3)_3COOH$	$CH_3SC_6H_{11}$	CH_3OH	2.2×10^{-4}	13.5	-33	e
HSO_5^-	Br^-	H_2O	1.0	10.0	-25	f
HSO_5^-	Cl^-	H_2O	1.8×10^{-3}	14.0	-24	f
CH_3CO_3H	Br^-	H_2O	0.26	13.1	-17	f
$C_6H_5CO_3H$	$S(CH_2C_6H_5Cl)_2$	$2\text{-}C_3H_7OH$	0.41^h	9.1	-22	g

[a] H. A. Liebhafsky and A. Mohammed, *J. Am. Chem. Soc.*, **55**, 3977 (1933); A. Mohammed and H. A. Liebhafsky, *J. Am. Chem. Soc.*, **56**, 1680 (1934).

[b] I. R. Wilson and G. M. Harris, *J. Am. Chem. Soc.*, **82**, 4515 (1960); I. R. Wilson and G. M. Harris, *J. Am. Chem. Soc.*, **83**, 286 (1961).

[c] C. G. Overberger and R. W. Cummins, *J. Am. Chem. Soc.*, **75**, 4783 (1953).

[d] J. O. Edwards and D. H. Fortnum, *J. Org. Chem.*, **27**, 407 (1962).

[e] L. Bateman and K. R. Hargrave, *Proc. Roy. Soc. (London)*, **A224**, 389, 399 (1954).

[f] D. H. Fortnum, C. J. Battaglia, S. R. Cohen, and J. O. Edwards, *J. Am. Chem. Soc.*, **82**, 778 (1960).

[g] C. G. Overberger and R. W. Cummins, *J. Am. Chem. Soc.*, **75**, 4250 (1953).

[h] At -30°C.

and iodide ions, thiocyanate ion, thiosulfate ion, sulfite ion, nitrite ion, amines (such as trialkylamines, anilines, and hydroxylamines), and organic sulfur compounds (such as dialkylsulfides and mercaptans).

The rate law [Eq. (5-1)] shows that the rate-determining step involves both nucleophile and peroxide, and the entropy of activation values suggest a precisely orientated complex. Therefore, the activated complex might be pictured as in (I), with the nucleophile coming in along the line defined by the two oxygen atoms. Evidence that direct oxygen transfer takes place has been

$$R \diagdown O-O \cdots N \diagdown R'$$

(I)

found in a variety of cases; one is the oxidation of nitrite ion by peroxyacids. It was observed[4] that the nitrate ion formed in the reaction contains one oxygen atom (as shown by isotopic label) derived from the peroxide. Although the enthalpies of activation appear low when one considers that the nucleophile is coming into an electron-rich site, it seems certain from the large discrimination between nucleophiles that bond formation is significant in the transition state.

In the activated complex, breaking of the oxygen-oxygen bond must be in process in order that the number of electrons on the outer peroxide oxygen does not exceed the number permitted by the Pauli exclusion principle. The group RO^- leaves with the extra electrons. Any stabilization of this leaving anion will make it less basic and will also enable the oxygen-oxygen bond to break more easily; therefore an inverse correlation between rate of oxidation and leaving-anion basicity may be predicted. Some data that confirm this are presented in Table 5-2. If one constructs a linear free-energy plot of the rate of oxidation of bromide ion by various monosubstituted peroxides against the basicity of the leaving anion, a good correlation is observed.

The order of nucleophilic strength of some reducing agents has been studied with a number of peroxides. Data obtained with hydrogen peroxide are shown in Table 5-3. Roughly, the order of strength for displacement on peroxide oxygen parallels that for

Table 5-2 *Rates of halide oxidations by peroxides[a]*

ROOH	pK_a(ROH)	$\log k(I^-)$	$\log k(Br^-)$	$\log k(Cl^-)$
$H_3O_2^+$	−1.7	$(5.8)^b$	$(2.9)^b$	$(0.4)^b$
HSO_5^-	2.0	very fast	0.0	−2.8
H_3PO_5	2.0	very fast	$(0.9)^c$	
CH_3CO_3H	4.7	very fast	−0.8	
$H_2PO_5^-$	7		−1.5	
HPO_5^-	12		< -3.0	
H_2O_2	15.7	−0.2	−4.6	−7.0
Cumene hydroperoxide	> 16	−3.2		

[a] At 25°C in aqueous solution; see Ref. 1.
[b] After correction for estimated K_a of $H_3O_2^+$; value used was $pK_a = -4.7$.
[c] After correction for estimated K_a of H_3PO_5; value used was $pK_a = 0.0$.

Table 5-3　*Hydrogen peroxide reactions with nucleophiles*[a]

Nucleophile	E_n	k_2, liter mole^{-1} sec^{-1}	k_3, liter^2 mole^{-2} sec^{-1}	Note
SO_3^-	2.57	2×10^{-1}		b
$S_2O_3^-$	2.52	2.5×10^{-2}	1.7	
Thiourea	2.18			c
I^-	2.06	6.0×10^{-1}	10.5	
$S(CH_2CH_2OH)_2$		2.2×10^{-3}	2.3×10^{-2}	
CN^-	2.02	1.0×10^{-3}		
$N(C_2H_5)_3$	2.0	3.3×10^{-4}		
SCN^-	1.83	5.2×10^{-4}	2.5×10^{-2}	
NO_2^-	1.73	3×10^{-7}		b
OH^-	1.65	$\leqslant 1 \times 10^{-7}$		d
Br^-	1.51	2.3×10^{-5}	1.4×10^{-2}	
Cl^-	1.24	1.1×10^{-7}	5.0×10^{-5}	
CH_3COO^-	0.95			e
SO_4^-	0.59			e
H_2O	0.00			d

[a] Data at 25°C in aqueous solution; see Ref. 1.
[b] Also evidence for a displacement through isotope-tracer studies.
[c] Too fast to measure.
[d] Isotope-exchange studies indicate rate too slow to measure.
[e] No evidence found for a displacement.

displacement on saturated carbon. This is not surprising since a saturated carbon atom and a peroxide oxygen are structurally similar. In addition to the two single covalent bonds, each peroxide oxygen has in the valence shell two orbitals with pairs of electrons. In so far as an incoming nucleophile is concerned, these spare electron pairs on oxygen act as groups in the same sense as the hydrogens on a methyl carbon; the hydrogens are merely protons buried in an orbital containing a pair of electrons. The exact nature of the bonding and electron distribution on oxygens is still unsettled, but present data can be explained on the basis that the four orbitals in the second major quantum level form a set of sp^3 hybrid orbitals. Then, since peroxide oxygen and alkyl carbon are stereochemically analogous and are electronically saturated atoms in the same row on the periodic table, the observation that the orders of nucleophilic strength are similar seems reasonable.

There are, nevertheless, some differences in these nucleophilic orders. Hydroxide ion, which is strong in nucleophilic displacements on carbon, is weak in the peroxide case, and attempts[5, 6] to measure the rate of the process

$$HOOH + O*H^- \rightarrow HOO*H + OH^-$$

using labeled oxygen have been unsuccessful because it is extremely slow. By way of contrast, olefins, which are good nucleophiles in displacement on peroxide oxygen (in epoxide formation),[3] are poor in the case of saturated carbon. There is also a difference in the quantitative aspects of the orders. Iodide ion is about one-million-fold more reactive than chloride ion in displacements on peroxide oxygen, yet these two halide ions differ by only a factor of 250 in rates on saturated carbon.

Two factors must enter into any explanation of these differences. Since oxygen is more electronegative than carbon, it is more difficult to remove from it an electron pair such as that on a leaving anion and bond breaking in a nucleophilic displacement step is thus retarded in the oxygen transition state. Bond formation to the incoming nucleophile must be in process before breaking of the peroxide bond can be highly developed. This is reflected in the greater spread of rates observed in displacements on oxygen as compared to displacements on carbon.

The second factor is the electron-electron repulsion (resulting from the Pauli exclusion principle) that occurs in the transition state. In the nucleophilic displacement reaction, as in most reactions, a large portion of the activation energy is required to overcome the van der Waals forces between nonbonded atoms in the transition state. The peroxide oxygen, like the alkyl carbon, is electronically saturated but, because of its higher electronegativity, is less polarized. Therefore, it is more difficult for a nucleophile with its pair of electrons to approach the oxygen. The difficulty will be marked in cases of nucleophiles such as hydroxide ion, which have several nonbonded pairs of electrons in the valence shell. This repulsion between the electrons on the incoming oxygen nucleophile and those on the peroxide oxygen is the reason, presumably, why oxygen-containing nucleophiles are unreactive in displacements on oxygen. If, on the other hand, the nucleophile has unfilled low-energy orbitals that can accept pairs of electrons, a certain amount of the electron-electron repulsion between nucleophile and peroxide oxygen is alleviated; the oxygen electrons use the nucleophile orbitals to form a pi bond in the transition state. As discussed previously, a relationship between unfilled orbitals and nucleophile polarizability seems to exist.[7]

Because of the nonbonded pairs of electrons on the oxygens, peroxides may be expected to show acid catalysis in their reactions with nucleophiles. As Ross has pointed out,[8] it should be easier for $H_3O_2^+$ than for H_2O_2 to transfer an OH^+ to a nucleophile. The acid catalysis normally appears as a second term in the rate law; the kinetics over a range of pH values give the law

$$v = k_2[H_2O_2][N] + k_3[H_2O_2][N][H^+] \tag{5-2}$$

Therefore there are two activated complexes that differ from each other in the number of hydrogen ions they have. Some of these reactions, with rate constants for both terms, are listed in Table 5-3; it is worth noting that *tert*-butyl hydroperoxide oxidizes nucleophiles by a rate law identical to Eq. (5-2). The catalysis is known to be of the specific-acid type, since general acids such as acetic acid are not catalysts. Further, the catalysis cannot be due to the formation of the electron-deficient ion OH^+ since the reaction

$$H_2O_2 + H^+ \rightleftharpoons H_2O + OH^+$$

would have to reach equilibrium more rapidly than the nucleophiles are oxidized. Such certainly is not the case, for the rate of oxygen exchange with water is too slow to measure.

Of several possible mechanisms one fits the rate law and the numerical value of k_3 better than the rest. If the proton is added to hydrogen peroxide to form $H_3O_2^+$ in a rapid equilibrium, the second step can be a rate-determining displacement on oxygen with water as the leaving group. The steps are

$$H_2O_2 + H^+ \rightleftharpoons H_3O_2^+ \qquad \text{fast}$$
$$H_3O_2^+ + N \rightarrow H_2O + HON^+ \qquad \text{slow}$$

Since the rate step is a displacement on a substituted peroxide, it should be possible to compare its rate with those for other cases in Table 5-2. With the estimated[1] value of -4.7 for the pK_a of $H_3O_2^+$, the k_3 rate constants have been recalculated and included as the top line of data in Table 5-2; the calculated values are in good agreement with expectation for a monosubstituted peroxide with water as a leaving group.

Although these data are rather conclusive in indicating that the reactions of hydrogen peroxide in aqueous solution are subject to specific-acid catalysis, some cases of general-acid catalysis of peroxide reactions are also known. Certain acidic solvents seem to be important for facilitating the reactions of peroxides with nucleophiles.

The manner in which peroxide reaction rates depend on solvent nature suggests that significant charge separation does not occur in the transition state. If charge separation occurred, the rate constant would be a marked function of the dielectric constant and the entropy of activation would become more negative as the solvent polarity decreased.[9] As may be seen in Table 5-4, neither condition is fulfilled in the recently studied oxidation of thioxane by hydrogen peroxide[10]:

$$H_2O_2 + S\overset{CH_2-CH_2}{\underset{CH_2-CH_2}{\diagdown}}O \rightarrow H_2O + O-S\overset{CH_2-CH_2}{\underset{CH_2-CH_2}{\diagdown}}O$$

The rate constants show that the reaction proceeds fastest in the solvent acetic acid, which has a low dielectric constant ϵ and a low

Table 5-4 *Solvent effects on peroxide oxidation of thioxane[a]*

Solvent[b]	ϵ^c	Z^d	$k_2(25°C)$, liter mole^{-1} sec^{-1}	$\Delta H^{\ddagger}$	$\Delta S^{\ddagger}$
CH_3CO_2H	6.15	79.2	2.4×10^{-2}	10.5	-33
H_2O	78.5	94.6	2.6×10^{-3}	13.0	-27
D_2O^e	78.3		1.5×10^{-3}	12.9	-28
EG	37.0	85.1	5.0×10^{-4}	13.6	-28
CH_3OH	32.6	83.6	6.2×10^{-5}	14.5	-29
$(CH_3)_2CHOH$	18.3	76.3	1.1×10^{-5}	15.3	-30
$(CH_3)_3COH$	10.9		5.5×10^{-6}	17.4	-24
NMA	178.9		2.2×10^{-6}	18.7	-22
Dioxane[f]	2.209	75.4	$<7 \times 10^{-7}$		
DMF	37.8	68.5	$<1 \times 10^{-7}$		

[a] Rate data from Ref. 10.

[b] Solvent symbols: ethylene glycol, EG; N-methylacetamide, NMA; dimethyl formamide, DMF.

[c] Values from A. A. Maryott and E. R. Smith, *Natl. Bur. Std. (U.S.) Circ.*, **514** (1951).

[d] Values from E. M. Kosower, *J. Am. Chem. Soc.*, **80**, 3253 (1958).

[e] The isotope effect (k_{HOH}/k_{DOD}) is 1.68 at 25°C.

[f] Mechanism is different (third-order kinetics) in dioxane.

solvent-polarity-response parameter Z yet is an acidic molecule. Water follows, with the alcohols next in order of decreasing acidity. In the three solvents without oxygen-hydrogen bonds the rates are lowest, despite high dielectric constants in some cases. Also, probably as a result of the formation of water, the rate increases with time in these last solvents. It had been suggested[11] on the basis of several pieces of evidence that a cyclic activated complex of the type (II), where

$$\begin{array}{c} R \\ \diagdown \\ O\text{—}O \diagdown \diagdown H \\ \vdots \qquad \diagdown H \\ H\text{—}X \end{array}$$

(II)

HX is an *acidic* solvent molecule, is obtained in certain oxidations on nucleophiles by hydroperoxides. In agreement with this activated complex, a deuterium-isotope effect has been observed in aqueous solution, and the enthalpy of activation increases as rate decreases.[10] All the observations are in agreement with an activated complex involving proton transfers assisted by solvent.

5-2 PEROXYANIONS AS NUCLEOPHILES

One of the surprising observations in the field of peroxide mechanisms is the high nucleophilic power of the peroxyanions.[12] Some data on rates for nucleophilic attack by hydroxide ion and by perhydroxide ion are given in Table 5-5. In spite of the fact that the hydroxide ion is ten-thousand-fold more basic, it is much weaker as a nucleophile. Although basicity alone is not a general index of nucleophilic strength, basicity is usually found to be a good index whenever the nucleophiles have a common attacking atom. In the present case, oxygen is the attacking atom common to both hydroxide ion and perhydroxide ion; therefore, the latter ion would be expected to be the weaker nucleophile.

It is hard to understand why this inversion takes place. Although the binding strengths of the two ions to Lewis acids may not correspond exactly to their binding strengths to the proton, it is doubtful that anything close to a complete inversion takes place. Therefore, the high nucleophilic reactivity of the perhydroxide ion as compared to the hydroxide ion would seem to be a transition-state effect.

It may be that, in the transition state for a nucleophilic displacement reaction, the perhydroxide ion has some unusual

Table 5-5 *Relative rates for attack by perhydroxide and hydroxide ion[a]*

Substrate	k_{OOH^-}, liter mole^{-1} sec^{-1}	k_{OH^-}, liter mole^{-1} sec^{-1}	Rate ratio[b]
p-Nitrophenyl acetate[e]	3×10^4	15	2000
Sarin[d]	1.5×10^3	25	60
Paraoxon[e]	5.8×10^{-1}	1×10^{-2}	70
Benzonitrile[f]	7.3	1.14×10^{-4}	70,000
$Si(aca)_3^{+g}$	8×10^4	1.7×10^3	50
Benzyl bromide[h]	5.4×10^{-2}	1.6×10^{-3}	35

[a] Most of these rate constants are for 25°C in aqueous solution.
[b] Compare to basicity ratio $(K_{OOH^-}/K_{OH^-} = 0.0001)$.
[e] Ref. 18.
[d] Ref. 17.
[e] J. Epstein, M. M. Demek, and D. H. Rosenblatt, *J. Org. Chem.*, **21**, 796 (1956).
[f] K. B. Wiberg, *J. Am. Chem. Soc.*, **77**, 2519 (1955).
[g] R. G. Pearson, D. N. Edgington, and F. Basolo, *J. Am. Chem. Soc.*, **84**, 3233 (1962).
[h] R. G. Pearson and D. N. Edgington, *J. Am. Chem. Soc.*, **84**, 4607 (1962).

binding, a possibility that has recently been discussed.[7] The unusual binding must be an electronic effect rather than a cyclic transition state stabilized by hydrogen bonding; this conclusion is based on the observation[18] that the ion CH_3OO^- is as strong a nucleophile as HOO^-.

5-3 PEROXYANIONS AS LEAVING GROUPS

For a simple S_N2 reaction, the reverse process is also a nucleophilic displacement. What in the forward reaction is a nucleophile becomes a leaving group in the reverse reaction. Leaving-group character can be defined in terms of relative rates of loss of donor particles from similar substrates under attack by a common nucleophile, as in the reaction

$$N + RX_n \rightarrow NR + X_n$$

where X_n is a series of leaving groups. Since there are few comparative data for rates as a function of X_n, there are no established scales for leaving-group character. For reasons discussed in Section

5-1, however, it would be expected that, at least for those leaving groups with a common atom, the rate of release of X_n should increase as the basicity of X_n decreases.

The question of leaving-group character is relevant to the mechanisms of peroxide reactions since, by microscopic reversibility, a good nucleophile should also be a good leaving group provided the binding strength is normal. Peroxyanions are found to be highly nucleophilic, yet they do not appear to show exceptional ground-state binding; thus it is predicted that peroxyanions should be good leaving groups. Some data that bear this out are given in Table 5-6, which shows the susceptibility of peroxy compounds to hydrolysis. The two peroxyphosphates hydrolyze faster than the other compounds, including the p-nitrophenyl phosphate, which, on the grounds of leaving-group basicity, would be expected to hydrolyze more rapidly than peroxymonophosphoric acid.

It is again difficult to give a reason for the high reactivity, but several facts are significant. The hydrolysis of the peroxyphosphates, as in the other cases in this table, involves phosphorus-oxygen bond cleavage as demonstrated by oxygen-isotope experiments. The hydrolysis of peroxydiphosphate is about fiftyfold faster than that of peroxymonophosphate; a similar rate difference has been observed in the case of the analogous peroxysulfates. This difference is understandable, since the OOH^- ion is about two-hundred-fold more basic than the ion $SO_5^=$. Another interesting observation is that fluoride ion is a nucleophilic catalyst for the hydrolysis of peroxydiphosphates. These data suggest that

Table 5-6 *Rates for acid hydrolysis of phosphate compounds*[a]

Compound	Leaving group	Rate[b], sec^{-1}
H_2O_3POH	OH^-	$6 \times 10^{-7}(100°)$
$H_2O_3POCH_3$	OCH_3^-	$1 \times 10^{-6}(100°)$
$H_2O_3POC_6H_5$	$OC_6H_5^-$	$\simeq 4 \times 10^{-6}(100°)$
$H_2O_3POC_6H_4NO_2$	$OC_6H_4NO_2^-$	$\simeq 8 \times 10^{-5}(73°)$
H_2O_3POOH	OOH^-	$8 \times 10^{-5}(61°)$
$H_2O_3POOPO_3H_2$	$OOPO_3H_2^-$	$\simeq 6 \times 10^{-3}(63°)$

[a] All cases involve P—O bond cleavage; data collected in Ph.D. thesis of C. J. Battaglia, Brown University, 1962.

[b] In 1 M $HClO_4$; temperature in parentheses.

the hydrolysis reaction is a nucleophilic displacement by water on phosphorus.

The high nucleophilic reactivity of peroxyanions and the rapid hydrolysis of peroxy compounds are undoubtedly related, since both reactions appear to proceed by a nucleophilic displacement mechanism. It is concluded that the OOR$^-$ group allows the reaction to proceed in both directions by a path with an unusually low free energy of activation. The reason for this is not understood beyond the fact that it is tied in with the alpha effect discussed in Section 4-3.

5-4 PEROXIDE DECOMPOSITION

One of the most interesting of the nonradical reactions is the decomposition observed for several peroxides of the type ROOH:

$$2ROOH \rightarrow 2ROH + O_2$$

Peroxide decompositions are quite susceptible to trace-metal catalysis,[13-15] but addition of a strong sequestering agent has been found to inactivate the metal ions (in most but not all cases) so that the true decomposition can be studied. Caro's acid,[13] peroxybenzoic acids,[14] peroxymonophosphoric acid,[15] peroxyacetic acid,[15] and peroxychloroacetic acid[15] have all been found to show a similar behavior in their kinetic laws for decomposition. At constant pH, the rate is second-order in peroxide concentration. The pH profile for Caro's acid is shown in Figure 5-1. At pH = pK_a (for the peroxide proton) the plot shows a maximum; on the low-pH side of this maximum the rate is first-order in hydroxide ion concentration, and on the high-pH side of the maximum the rate is first-order in hydrogen ion concentration. When these facts are put together, the rate law can be reduced to the form of Eq. (5-3).

$$v = k[ROOH][ROO^-] \tag{5-3}$$

This rate has been observed with each of the peroxides listed above. Although the rate constants for these peroxides as given in Table 5-7 show some variation, the fact that the rate laws are identical suggests (incorrectly, as will be seen!) that the mechanism does not significantly depend on the nature of the R group. The activation parameters for the decomposition of Caro's acid were found to be 12.4 kcal mole^{-1} and -20.1 mole^{-1} deg^{-1} for $\Delta H^{\ddagger}$ and $\Delta S^{\ddagger}$, respectively. The similarity of these values to

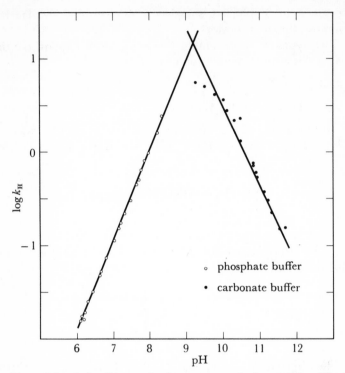

Figure 5-1   *A plot of the second-order rate constant for decomposition of Caro's acid (peroxymonosulfuric acid) as a function of pH. Note that the pH at intersection of the two lines corresponds to the known pK_a for HSO_5^-.*

Table 5-7   *Rates of peroxide decompositions for second-order, base-catalyzed path[a]*

Peroxide	$pK_a(ROH)$	$pK_a(ROOH)$	k, liter mole^{-1} sec^{-1}	Ref.
HSO_5^-	2.0	9.4	0.39	13
$ClCH_2CO_3H$	2.8	7.2	0.4(15°)	15
$C_6H_5CO_3H$	4.0	7.8	9×10^{-3}	14
CH_3CO_3H	4.8	8.2	1.1×10^{-2}	15
HPO_5^-	12	12.5	$1.9 \times 10^{-3}(36°)$	15

[a]Rate law is $k[ROOH][ROO^-]$; data are for 25°C (except as noted) and for aqueous solutions.

those given in Table 5-1 suggests a nucleophilic displacement mechanism for the rate step.

Two types of displacement mechanism can explain the data. The first[13] is a displacement on oxygen by oxygen

$$HSO_5^- + SO_5^- \rightarrow HSO_6^- + SO_4^-$$

$$HSO_6^- \rightarrow H^+ + O_2 + SO_4^-$$

The initial step, which is postulated to be rate-determining, involves an activated complex (III) that forms an intermediate

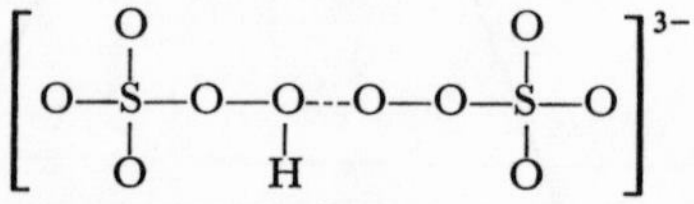

$$\left[O-\overset{\overset{O}{\|}}{\underset{\underset{O}{|}}{S}}-O-O\cdots O-O-\overset{\overset{O}{\|}}{\underset{\underset{O}{|}}{S}}-O \right]^{3-}$$
$$ H$$

(III)

(IV). The intermediate could be expected to break down rapidly to give products.

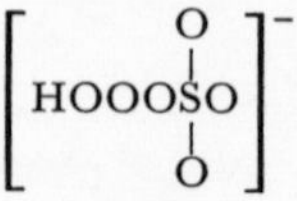

$$\left[HOOO\overset{\overset{O}{|}}{\underset{\underset{O}{|}}{S}}O \right]^{-}$$

(IV)

The alternative mechanism[14] differs in that the site of attack by the peroxyanion is the central atom of the R group. In the activated complex, using peroxyacetic acid (V) as an example, the

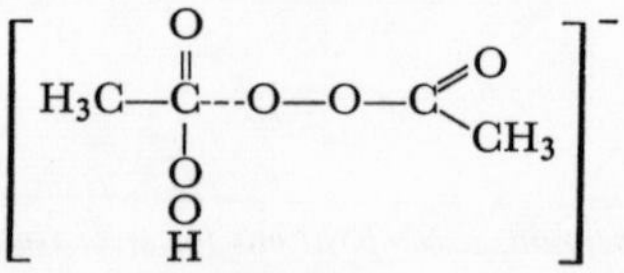

$$\left[H_3C-\overset{\overset{O}{\|}}{\underset{\underset{\underset{\underset{H}{|}}{\overset{|}{O}}}{\overset{|}{O}}}{C}}\cdots O-O-C\overset{\diagup O}{\diagdown CH_3} \right]^{-}$$

(V)

incoming peroxyanion becomes attached to the carbonyl carbon to form an intermediate (VI), which breaks down to form products by separation at the bonds designated by arrows.

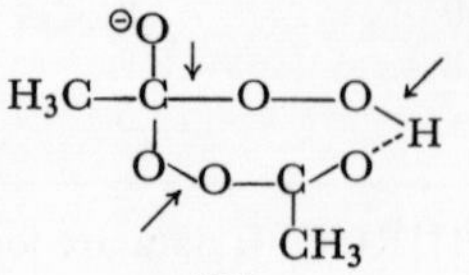

$$H_3C-\overset{\overset{\ominus O}{|}}{C}-O-O\diagdown_{H} \atop \underset{O}{\diagdown}O-C\overset{O}{\diagup} \atop CH_3$$

(VI)

Fortunately, these two mechanisms can be distinguished by the isotope-tracer technique. From a mixture of double-labeled peroxyacetic acid (VII) and nonlabeled peroxyacid, the first

$$H_3C-\overset{\overset{\displaystyle O}{\|}}{C}\diagdown_{O*-O*-H}$$

(VII)

mechanism predicts that the product oxygen molecules should be predominantly scrambled (to give O_2 of mass 34). The second mechanism predicts little or no scrambling; the product oxygen gas should be predominantly a mixture of normal (mass 32) and double-labeled (mass 36) species. Recent measurements[15] show that little scrambling ($\simeq 17$ per cent) occurs; therefore the second mechanism is strongly favored. The minor degree of scrambling observed does not necessarily denote attack at oxygen, for labeled oxygen molecules can be scrambled by solutions containing small amounts of the radical O^-.

Since both sulfur and phosphorus have open d orbitals available for attack by the strongly nucleophilic peroxyanion, the second mechanism is also feasible for these inorganic peroxyacids.† Since displacements on oxygen by oxygen nucleophiles are very slow (if even existent), as was mentioned in Chapter 4, the tracer-isotope results are in good agreement with prior data.

These results predict that an R group that cannot expand its coordination number will, of necessity, force the peroxide to take an alternative path for decomposition. The decompositions of tert-alkyl hydroperoxides are exceedingly slow in aqueous base[15, 16] even though the rate of decomposition first increases and then decreases as the concentration of base increases. Similarly, the alkaline decomposition of hydrogen peroxide goes through a rate maximum[17]; however, there is good evidence that this decomposition depends on the presence of trace metals.[15] It is evident that the noncatalyzed decomposition of hydrogen peroxide is a very slow process.

---

† It has now been found (by E. Koubek, G. Levey, and J. O. Edwards, *Inorg. Chem.*, **3**, 1331 (1964)) that the decomposition of doubly labeled Caro's acid in the presence of normal Caro's acid gives oxygen that is about 91 per cent scrambled. This result is consistent with the mechanism involving oxygen attack at oxygen.

5-5 HYPOHALITE REACTIONS

Some of the hypohalous acid reactions are similar to peroxide reactions in that an oxygen atom is transferred to a reducing substrate. For example, the oxidation of nitrite ion by hypochlorous or hypobromous acid gives nitrate ion with direct oxygen-atom transfer as shown by isotope studies.[18] The rate law has been found[19] to be

$$v = k[\text{HOCl}][\text{NO}_2^-] \tag{5-4}$$

which is analogous to the rate law for peroxide oxidation of nitrite ion.[20] The data on this and several other oxidations by hypochlorous acid, as collected by Lister,[19] are suggestive of a mechanism involving nucleophilic displacement by the reducing agent on the oxygen.

Another behavior that is analogous to that of a peroxide is the high nucleophilic strength of the anion. Hypochlorite ion has been found to react as a nucleophile to substrates such as tetrahedral phosphorus compounds[21] and p-nitrophenyl acetate.[22]

Other nonradical hypohalite reactions are discussed in the interesting paper by Taube[23]; also, some exchange reactions of hypochlorite and hypobromite ions are covered in Chapter 8 of this book.

It is worth noting that hypochlorous acid reactions are subject to trace-metal catalysis and can proceed by radical mechanisms; this is another similarity to peroxide reactions.

References

1. J. O. Edwards, in J. O. Edwards (ed.), *Peroxide Reaction Mechanisms*, Wiley-Interscience, New York, 1962, pp. 67–106.
2. W. K. Wilmarth and H. Haim, in J. O. Edwards (ed.), *Peroxide Reaction Mechanisms*, Wiley-Interscience, New York, 1962, pp. 175–226.
3. D. Swern, *Chem. Rev.*, **45**, 1 (1949).
4. M. Anbar and H. Taube, *J. Am. Chem. Soc.*, **76**, 6243 (1954).
5. M. C. R. Symons, *Chem. & Ind.* (*London*), **1960** (No. 48), 1480.
6. M. Anbar, *J. Am. Chem. Soc.*, **83**, 2031 (1961).
7. J. O. Edwards and R. G. Pearson, *J. Am. Chem. Soc.*, **84**, 16 (1962).
8. S. D. Ross, *J. Am. Chem. Soc.*, **68**, 1484 (1946).
9. A. A. Frost and R. G. Pearson, *Kinetics and Mechanism*, 2d ed., Wiley, New York, 1961, p. 122ff.
10. M. Dankleff, Ph.D. thesis, Brown University, 1963.

11. L. Bateman and K. R. Hargrave, *Proc. Roy. Soc. (London)*, **A224**, 389, 399 (1954).

12. C. A. Bunton, in J. O. Edwards (ed.), *Peroxide Reaction Mechanisms*, Wiley-Interscience, New York, 1962, pp. 11–28.

13. D. L. Ball and J. O. Edwards, *J. Am. Chem. Soc.*, **78**, 1125 (1956); J. F. Goodman and P. Robson, *J. Chem. Soc.*, **1963**, 2871.

14. J. F. Goodman, P. Robson, and E. R. Wilson, *Trans. Faraday Soc.*, **58**, 1846 (1962).

15. E. Koubek, M. L. Haggett, C. J. Battaglia, K. M. Ibne-Rasa, H. Y. Pyun, and J. O. Edwards, *J. Am. Chem. Soc.*, **85**, 2263 (1963).

16. M. S. Kharasch, A. Fono, W. Nudenberg, and B. Bischof, *J. Org. Chem.*, **17**, 207 (1952).

17. F. R. Duke and T. W. Haas, *J. Phys. Chem.*, **65**, 304 (1961).

18. M. Anbar and H. Taube, *J. Am. Chem. Soc.*, **80**, 1073 (1958).

19. M. W. Lister and P. Rosenblum, *Can. J. Chem.*, **39**, 1645 (1961).

20. J. O. Edwards and J. J. Mueller, *Inorg. Chem.*, **1**, 696 (1962).

21. J. Epstein, V. E. Bauer, M. Saxe, and M. M. Demek, *J. Am. Chem. Soc.*, **78**, 4068 (1956); L. Larsson, *Svensk Kem. Tidskr.*, **70**, 405 (1958).

22. W. P. Jencks and J. Carriuolo, *J. Am. Chem. Soc.*, **82**, 1778 (1960).

23. H. Taube, *Record Chem. Progr.*, **17**, 25 (1956).

6

Replacements in Octahedral
Complexes

Some of the best-known inorganic compounds are those in which a central metal ion is surrounded by six ligands in octahedral configuration. The replacement of one ligand by another has been studied intensively and extensively; yet the mechanism by which this replacement occurs has not been completely unraveled at this writing. At least in part, the difficulty lies in our lack of understanding of the nature of bonding and electron distribution in such complexes. It will, therefore, be necessary to mention bonding theory along with kinetics in this chapter. For an excellent and detailed review on these subjects up to 1958, the reader is referred to the book by Basolo and Pearson.[1] There are also two recent reviews on kinetics and mechanisms of reactions in coordination compounds.[2, 3]

6-1 LABILE AND INERT COMPLEXES

Taube[4] has pointed out that rates of replacement in octahedral transition metal complexes fall into two rough categories. The classification *labile* is applied to very reactive complexes, that is, those whose reactions are complete within the time required for mixture (1 min, room temperature, about 0.1 M solutions); the classification *inert* is applied to those that react at rates too slow to measure or at rates that can be followed by conventional techniques.

The term *inert* should be used only in the comparison of rates and should not be confused with the term *stable*, which has a thermodynamic meaning. Although an inert complex may be stable, the two properties are not necessarily related. A stable complex can be labile; for example, $[Hg(CN)_4]^=$ has the very high formation constant of 10^{42}, yet in solution it exchanges cyanide ligands with labeled cyanide ion at a very fast rate. On the other hand, a thermodynamically unstable complex such as $[Co(NH_3)_6]^{3+}$ can remain unchanged in acid solution for weeks.

The striking thing about the rates of the transition metal complexes is that they correlate with electronic configuration as indicated by magnetic susceptibility of the complex. A qualitative treatment[4] of the data in terms of valence-bond theory[5] has been successful, as may be seen in Table 6-1. Here complexes that are considered to be covalent, with d^2sp^3 hybridization, are divided into the two rate classes. In the labile complexes, there is at least one d orbital available for electron addition from an entering ligand, whereas in the inert complexes every d orbital contains at least one electron.

An explanation is available for this interesting correlation, if it is assumed that the reaction involves an S_N2 mechanism. Electrons from the entering ligand occupy the open d orbital to form a seven-coordinate intermediate, which then releases the ligand being replaced and forms a new octahedral complex. The formation of the seven-coordinate intermediate is facilitated (fast-rate) by an empty inner d orbital. If, however, the inner d orbitals are occupied, the incoming ligand must employ the less stable d orbitals of higher principal quantum number; in this case slow rates would be expected because the entering ligand cannot bond as strongly as the six ligands already present.

Transition metal complexes of the outer-orbital type, that is, those with sp^3d^2 hybrid bonding, are generally labile.[4] For example, the octahedral complexes of "spin-free" Mn(II), Fe(II), Fe(III), Co(II), Ni(II), Cu(II), and Cr(II) all exchange ligands rapidly. In terms of valence-bond theory, this observation can be correlated

Table 6-1   *Inner-orbital six-coordinate complexes*

Electronic configuration			Central metal ions
d	s	p	
Labile complexes			
○○○⊙⊙	⊙	⊙⊙⊙[a]	Sc(III), Y(III), rare earths(III), Ti(IV), Zr(IV), Hf(IV), Ce(IV), Th(IV), Nb(V), Ta(V), Mo(VI), W(VI)
⊙○○⊙⊙	⊙	⊙⊙⊙[a]	Ti(III), V(IV), Mo(V), W(V), Re(VI)
⊙⊙○⊙⊙	⊙	⊙⊙⊙[a]	Ti(II), V(III), Nb(III), Ta(III), Mo(IV), W(IV), Re(V), Ru(VI)
Inert complexes			
⊙⊙⊙⊙⊙	⊙	⊙⊙⊙[a]	V(II), Cr(III), Mo(III), W(III), Mn(IV), Re(IV)
⊙⊙⊙⊙⊙	⊙	⊙⊙⊙	$[Cr(CN)_6]^{4-}$, $[Cr(bipy)_3]^{++}$, $[Mn(CN)_6]^{3-}$, Re(III), Ru(IV), Os(IV)
⊙⊙⊙⊙⊙	⊙	⊙⊙⊙	$[Cr(bipy)_3]^{+}$, $[Mn(CN)_6]^{4-}$, Re(II), $[Fe(CN)_6]^{3-}$, $[Fe(Ph)_3]^{3+}$, $[Fe(bipy)_3]^{3+}$, Ru(III), Os(III), Ir(IV)
⊙⊙⊙⊙⊙	⊙	⊙⊙⊙	$[Mn(CN)_6]^{5-}$, $[Fe(CN)_6]^{4-}$, $[Fe(Ph)_3]^{++}$, $[Fe(bipy)_3]^{++}$, Ru(II), Os(II), Co(III) (except CoF_6^{3-}), Rh(III), Ir(III), Ni(IV), Pd(IV), Pt(IV)

[a] In these cases the electronic configuration is assumed.

with the weakness of bonds of the sp^3d^2 type as compared with d^2sp^3 bonds.

For octahedral complexes of the non-transition metals, inertness is not related to electronic configuration but is apparently correlated with high charge on the central metal ion. For example, PF_6^- is inert, whereas AlF_6^{3-} is labile. Other inert complexes are SF_6, $SbCl_6^-$, and AsF_6^-. As the charge on the central atom increases, the ligands already present are attracted more firmly and become more difficult to dislodge.

6-2   CRYSTAL-FIELD STABILIZATION

In recent years, a semiquantitative explanation of the rate differences between various complexes has appeared. Certain transition metal complexes are stabilized by placing the non-bonding electrons in d orbitals spatially distributed away from the ligands. The five d orbitals are not spatially equivalent; they have definite orientations relative to the geometric configuration of the ligands. In an octahedron, three of the d orbitals (d_{xy}, d_{yz}, d_{xz}) are directed between the ligands and therefore contain electrons of lower energy than those in the two orbitals ($d_{x^2-y^2}$, d_{z^2}) directed toward the ligands.

The *crystal-field stabilization energy* (*CFSE*), gained as a result of placing electrons preferentially in the orbitals directed away from the ligands, can be significant. It is particularly important to the present topic, since it is one factor in determining activation energy. Whether the octahedral complex with its attendant CFSE reacts by an S_N2 or an S_N1 mechanism, the symmetry is lowered and a decrease in CFSE usually occurs in going to the transition state. This loss of CFSE in the activated complex, expressed in units of the crystal-field-splitting parameter Dq, must be added to the other components of the activation energy and, as may be seen in Tables 6-2 and 6-3, can be sizable. The parameter Dq may be determined from the spacings of the spectral lines, and it changes with metal ion and ligand. It varies roughly between 700 and 3500 cm^{-1}; this corresponds to 2 to 12 kcal mole^{-1} per Dq unit. It must be remembered that there is one bond more or one bond less in the activated complex; therefore the bond strengths (also, the charges and electronegativities) will influence the activation energy in a manner independent of the Dq contribution. Only in chemically similar systems will the comparisons based on Dq values be valid; for example, the effect of charge is discussed in Section 6–7. In the case of $[Cr(OH_2)_6]^{3+}$, the CFSE of the ground state amounts to about 64 kcal mole^{-1}, of which from 12 to 24 kcal mole^{-1} can be lost in a transition state; this is equivalent to a decrease in rate of between nine and fourteen powers of ten!

In these tables the energy differences between the ground state and alternative transition states as computed by Basolo and Pearson[1] are presented. The values for ground states are for regular octahedra both in strong crystal fields (spin-paired complexes) and in weak fields (spin-free complexes), whereas those for transition states are for square pyramids (assumed for S_N1 mechanism) and for pentagonal bipyramids (assumed for S_N2 mechanism). The difference between the CFSE of the octahedron and the final CFSE is considered to be its contribution ΔE_a, to the

Table 6-2 *Crystal-field activation energies for dissociation mechanism*

Octahedral → square pyramid

	Strong fields			Weak fields		
System	Octahedral	Square pyramid	ΔE_a	Octahedral	Square pyramid	ΔE_a
d^0	$0Dq$	$0Dq$	$0Dq$	$0Dq$	$0Dq$	$0Dq$
d^1	4	4.57	-0.57	4	4.57	-0.57
d^2	8	9.14	-1.14	8	9.14	-1.14
d^3	12	10.00	2.00	12	10.00	2.00
d^4	16	14.57	1.43	6	9.14	-3.14
d^5	20	19.14	0.86	0	0	0
d^6	24	20.00˙	4.00	4	4.57	-0.57
d^7	18	19.14	-1.14	8	0.14	-1.14
d^8	12	10.00	2.00	12	10.00	2.00
d^9	6	9.14	-3.14	6	9.14	-3.14
d^{10}	0	0	0	0	0	0

Table 6-3 *Crystal-field activation energies for displacement mechanism*

Octahedral → pentagonal bipyramid

	Strong fields			Weak fields		
System	Octahedral	Pentagonal bipyramid	ΔE_a	Octahedral	Pentagonal bipyramid	ΔE_a
d^0	$0Dq$	$0Dq$	$0Dq$	$0Dq$	$0Dq$	$0Dq$
d^1	4	5.28	-1.28	4	5.28	-1.28
d^2	8	10.56	-2.56	8	10.56	-2.56
d^3	12	7.74	4.26	12	7.74	4.26
d^4	16	13.02	2.98	6	4.93	2.07
d^5	20	18.30	1.70	0	0	0
d^6	24	15.48	8.52	4	5.28	-1.28
d^7	18	12.66	5.34	8	10.56	-2.56
d^8	12	7.74	4.26	12	7.74	4.26
d^9	6	4.93	1.07	6	4.93	1.07
d^{10}	0	0	0	0	0	0

total activation energy of the reaction. A high value of ΔE_a implies a slow reaction, whereas a zero or negative value implies a relatively rapid reaction with no additional contribution to the normal activation energy.

For the strong-field complexes, which correspond to the inner-orbital complexes of Pauling,[5] it is seen that for the d^0, d^1, and d^2 systems there is no CFSE loss by either mechanism. These are known to be labile cases. For the d^3 and spin-paired d^4, d^5, and d^6 cases, there are positive ΔE_a values for both mechanisms; these are known[4] to be inert configurations. The spin-free complexes d^4, d^5, d^6, and d^7, which are termed outer-orbital in the valence-bond theory and weak-field complexes in the crystal-field theory, are predicted to be labile by both theories; observation has confirmed this prediction.

The single case for which the two theories give different predictions is that of the d^8 electronic configuration in an octahedral complex. The valence-bond treatment predicts a labile system, whereas the crystal-field treatment predicts that for any mechanism there should be a significant loss of CFSE in the transition state and therefore a low rate. This interesting case will be discussed in Section 6–7.

6-3 GENERAL ASPECTS

The factors that influence the rates and mechanisms of replacements in octahedral complexes are manifold, and the classification of complexes as inert or labile helps little in understanding the complete reaction mechanism. In this section other data relevant to mechanism will be presented, although few final conclusions can be reached. Since the rates of reactions involving Co(III) complexes are measurable and much data has been accumulated for the system, the examples used will be primarily Co(III) reactions.†

1. The replacement of one ligand by a second often proceeds by a two-step process

$$ML_5X + H_2O \rightarrow ML_5OH_2 + X$$

$$ML_5OH_2 + Y \rightarrow ML_5Y + H_2O$$

† Similarities and differences between Co(III) and Rh(III) hydrolysis mechanisms have been reported [S. A. Johnson, F. Basolo, and R. G. Pearson, *J. Am. Chem. Soc.*, **85**, 1741 (1963)].

in which the first step is replacement of the leaving ligand by solvent water (*aquation*) and the second step involves replacement of the coordinated water by the entering ligand (*anation*). Since most coordination-compound reactions are run in aqueous solution with water in vast excess, and since water is a good ligand in its own right, it is not surprising that such reactions follow the aquation-anation path. Stranks[2] points out that water is a poor solvent choice because it can interact in a variety of ways: (a) by primary solvation, as in $[Cr(OH_2)_6]^{3+}$; (b) by secondary solvation, as in the two weakly bound water molecules above and below the plane of a square-planar complex; (c) by hydrogen bonding; (d) by reacting as a nucleophile; and (e) by exerting its acid-base properties. All of these interactions are more or less important in coordination-compound reactions.

2. Such reactions are strongly base-catalyzed. Most nucleophiles are unreactive in comparison to the ever-present water; however, hydroxide ion is very reactive. For example, the hydrolysis of $[Co(NH_3)_4(OH_2)NO_3]^{++}$ follows first-order kinetics with an observed constant of the form

$$k_{obs} = k_1 + k_2[OH^-] \tag{6-1}$$

such that at pH 2 the contributions of the two paths are equal. Since the hydroxide ion is also the lyate ion, its role in the reaction mechanism is ambiguous, as will be seen in Section 6-5.

3. There is a strong rate dependence on the nature of the leaving ligand. For aquation of complexes of the type $[Co(NH_3)_5X]^{++}$, the order of reactivity of X^- is $HCO_3^- > NO_3^- > I^- > Br^- > H_2O > Cl^- > SO_4^= > F^- > CH_3COO^- > NCS^- > NO_2^-$. Also, for a series of substituted-acetato complexes, rates for both acid hydrolysis (aquation) and base-catalyzed hydrolysis decrease as the anion basicity increases; data to this effect are shown in Table 6-4. Apparently bond breaking is important in the transition state.

4. Steric effects slightly favor an intermediate of lower coordination number. The data in Table 6-4 show that the substituted acetates react in a manner suggesting that the bulk of the leaving group has no influence on the rate. Both trimethylacetate and trichloroacetate are bulky ligands; yet their rates correlate with dissociation constants in the same way as the rates for the less bulky ligands. This evidence is inconsistent with an S_N2 mechanism

Table 6-4 *Rates of acid and base hydrolyses[a] of some pentamineacetatocobalt(III) complexes, $[Co(NH_3)_5X]^{++}$*

X-	$K_a(HX)$, 25°C	k_{H_2O}, min⁻¹, 70°C	k_{OH^-}, liter mole⁻¹ min⁻¹, 25°C
CF_3COO^-	5×10^{-1}	3.3×10^{-3}	4.4
CCl_3COO^-	2×10^{-1}	3.2×10^{-3}	4.3
$CHCl_2COO^-$	5×10^{-2}	9.6×10^{-4}	1.6
CH_2ClCOO^-	1.4×10^{-3}	3.5×10^{-4}	2.5×10^{-1}
CH_2OHCOO^-	1.5×10^{-4}		7.0×10^{-2}
CH_3COO^-	1.8×10^{-5}	4.9×10^{-4}	4.2×10^{-2}
$CH_3CH_2COO^-$	1.5×10^{-5}	1.9×10^{-4}	2.7×10^{-2}
$(CH_3)_2CHCOO^-$	1.5×10^{-5}	1.6×10^{-4}	3.4×10^{-2}
$(CH_3)_3CCOO^-$	1.0×10^{-5}	2.6×10^{-4}	1.8×10^{-2}

[a] All data from F. Basolo, J. G. Bergmann, and R. G. Pearson, *J. Phys. Chem.*, **56**, 22 (1952).

involving water attack on the same side of the complex as the ligand.

In Table 6-5 data on the aquation of the first chlorine of some cobalt(III) complexes of the type *trans*-$[Co(AA)_2Cl_2]^+$, where AA is a chelating diamine ligand, are presented. With only one exception, the effect of increased bulk on the chelate ligand is to increase the rate of loss of chloride ion. This pattern certainly suggests an S_N1 mechanism. Steric crowding in the complex would be expected to inhibit an S_N2 reaction. When crowding is significant, on the other hand, an intermediate of lower coordination number such as might occur in a dissociative type of replacement mechanism could be formed at a more rapid rate.

5. The aquation of cobalt(III) chloride containing polydentate ligands is slower than that of analogous complexes containing monodentate ligands. The representative data given in Table 6-6 show that the rate decreases as the number of —CH_2CH_2— chelate links increases. In spite of its direction, this effect cannot be explained as steric hindrance to nucleophilic attack on the side of the complex *trans* to the leaving group. Such a hindrance would be vastly more serious than is observed.

A more plausible explanation lies in steric hindrance to

Table 6-5 *Rates of aquation[a] of trans-$[Co(AA)_2Cl_2]^+$*

$$[Co(AA)_2Cl_2]^+ + H_2O \rightarrow [Co(AA)_2OH_2Cl]^{++} + Cl^-$$

Diamine AA	$k \times 10^3$, min^{-1}
Ethylenediamine	1.9
Propylenediamine	3.7
dl-Butylenediamine	8.8
meso-Butylenediamine	250
Isobutylenediamine	130
Tetramethylethylenediamine	instantaneous
N-methylethylenediamine	1.0
N-ethylethylenediamine	3.6
N-propylethylenediamine	7.1

[a] Data from R. G. Pearson, C. R. Boston, and F. Basolo, *J. Am. Chem. Soc.*, **75**, 3089 (1953); rates in aqueous solution at pH 1 and 25°C.

solvation. The complex in both the ground state and the transition state must be solvated, with the solvation expected to be more significant in the transition state, where charge separation is occurring. Since the chelate groups would prevent effective solvation, the net effect should be a lowering of the rate as observed.

Table 6-6 *Rates of aquation of pentaminecobalt(III) chlorides[a]*

Ion[b]	No. of chelate links	$k \times 10^4$, min^{-1}
$[Co(NH_3)_5Cl]^{++}$	0	4.0
cis-$[Co(en)_2NH_3Cl]^{++}$	2	0.85
cis-$[Co(trien)NH_3Cl]^{++}$	3	0.40
$[Co(en)(dien)Cl]^{++}$	3	0.31
$[Co(tetraen)Cl]^{++}$	4	0.15

[a] Data (at pH 1 and 35°C) from R. G. Pearson, C. R. Boston, and F. Basolo, *J. Phys. Chem.*, **59**, 304 (1955).

[b] The symbols en, trien, dien, and tetraen represent ethylenediamine, triethylenetetramine, diethylenetriamine, and tetraethylenepentamine, respectively.

6. Acid-catalyzed hydrolysis also occurs with certain co-ordination compounds. Ligands that are basic or have a tendency to hydrogen-bond, such as carbonate in $[Co(NH_3)_5CO_3]^+$ or cyanide in $[Fe(CN)_6]^{4-}$, are susceptible to hydrolysis in acid solution. The postulated mechanism[6] is

$$[Co(en)_2F_2]^+ + H^+ \rightleftharpoons [(en)_2FCo—FH]^{++} \qquad\qquad \text{fast}$$

$$[(en)_2FCo—FH]^{++} + H_2O \rightarrow [(en)_2FCoOH_2]^{++} + HF \qquad \text{slow}$$

where the rate-determining step is loss of the HF molecule from the protonated complex. Since D_3O^+ is a stronger acid than H_3O^+, the fact that the rate in D_2O is almost twice as fast as the rate in H_2O supports this mechanism.

Other ligands that are susceptible to acid hydrolysis are the flexible chelates such as bipyridine, ethylenediamine, and EDTA. The postulated mechanism here involves primary dissociation from metal of one end of the ligand with concomitant protonation. For example, in the nickel-ethylenediamine complex,

$$(en)_2Ni^{++}\ \begin{array}{c} H_2 \\ {}^{/N}{}^{\diagdown}CH_2 \\ | \\ {}^{\diagdown}N{}^{\diagup}CH_2 \\ H_2 \end{array} \rightleftharpoons (en)_2Ni^{++}—NH_2CH_2CH_2NH_2 \xrightarrow{k_0} (en)_2Ni^{++} + en$$

$$H^+ \Big\Updownarrow$$

$$(en)_2Ni^{++}—NH_2CH_2CH_2NH_3^+ \xrightarrow{k_H} (en)_2Ni^{++} + enH^+$$

the proton provides an additional path for breakdown of the complex. Also, it seems reasonable that k_H should be greater than k_0.

7. To discover whether the cobalt(III) species in the transition state is a distinct intermediate having some stability and therefore reproducible selectivity, the oxygen-18 isotope discrimination in metal-catalyzed hydrolyses of $[Co(NH_3)_5X]^{++}$, where $X^- = Cl^-$, Br^-, and I^-, has been investigated by Posey and Taube.[7] The metal ions in order of their reactivity are $Hg^{++} > Ag^+ > Tl^{3+}$, and the reaction is the aquation

$$[Co(NH_3)_5X]^{++} + H_2O \xrightarrow{M} [Co(NH_3)_5H_2O]^{3+} + X^-$$

The results are presented in Table 6-7, where the values given are the ratios of the oxygen-18 content of the solvent to that of the

Table 6-7 *Fractionation effects in aquation of* $[Co(NH_3)_5X]^{++}$

	Metal catalyst		
Complex	Hg^{++}	Ag^+	Tl^{3+}
$[Co(NH_3)_5Cl]^{++}$	1.012	1.009	0.996
$[Co(NH_3)_5Br]^{++}$	1.012	1.007	0.993
$[Co(NH_3)_5I]^{++}$	1.012	1.010	1.003

product complex. The high values of the fractionation effects and the identity in size for the three complexes with Hg^{++} suggest that from all three reactants a common and distinct intermediate, presumably $[Co(NH_3)_5]^{3+}$, is formed, and that this intermediate has a reproducible preference for the oxygen-16 isotope in the water. The data for the less reactive Ag^+ and Tl^{3+} suggest that water participation in the transition state is significant and varies with each case, thus ruling out a distinct intermediate. Although in the presence of mercuric ion the mechanism of aquation is primarily one of complete bond breaking to form an intermediate, new bond formation is involved in the cases of the other two metal ions. Since some bond formation occurs even when the two metal ions aid the bond-breaking process, it is concluded that aquation in the absence of metal ions cannot be a true S_N1 reaction, for there should then be even greater solvent participation in the transition state.

8. Pressure effects on rate constants can, in principle, distinguish between an S_N1 and S_N2 reaction; in the former case the activated complex is larger than the reactant (the activation volume is positive), whereas the reverse is true for the latter case. The value of $\Delta V^{\ddagger}$ for the exchange reaction

$$[Co(NH_3)_5H_2O]^{3+} + H_2O^* \rightleftharpoons [Co(NH_3)_5H_2O^*]^{3+} + H_2O$$

has been found[8] to be $+1.2$, which though not large suggests an activated complex with more bond breaking than bond formation.

6-4 DISSOCIATION vs. DISPLACEMENT

Replacements in coordination compounds have been carefully studied with the hope of elucidating the mechanism, but as

Section 6-3 indicates, the results are not clear-cut. It seems appropriate at this point to jettison the notion of either a pure S_N1 or a pure S_N2 reaction for Co(III) complexes, and instead to view the mechanism in terms of the amount of bond formation occurring while bond breaking is taking place. No order of nucleophilic strength can be set up since the solvent water, because of its relatively high concentration, reacts to the essential exclusion of other ligands, with the questionable exception of hydroxide ion to be discussed below. Since, according to the data in Chapter 4, some nucleophilic order would be observed if bond formation as in an S_N2 process were significant, it can be concluded that little bond formation occurs in the transition state.

Most of the data given in Section 6-3 suggest that bond breaking is well developed in the transition state. The steric effects, the dependence of rate on leaving group, and the activation volume change all essentially agree that in the activated complex the bound ligand moves away from the metal ion with little help from an incoming group. On the other hand, the solvent-isotope-fractionation effects presented in Table 6-7 rule out a discrete, thermally equilibrated intermediate with a coordination number of five.

A mechanism consistent with these data involves extension of the ligand-to-metal bond almost to the point of breakage; the entering ligand then drops into the reactive site being opened. This mechanism predicts that the nearest nonbound ligand will drop into the open site; because of its high concentration this would usually be water. In reactions between cationic complexes and anionic ligands, however, ion pairing could hold the anion near the complex, making it easier for the anion to compete with water for the developing site. Such ion-pairing effects are known.[1, 2]

The relative amounts of bond breaking and bond making also depend on the bound but unreacting ligands on the cobalt. For example, a ligand that can release electrons to cobalt for formation of a pi bond with the developing open orbital will favor bond breaking, whereas an electron-withdrawing ligand will favor bond formation to the incoming nucleophile. In Table 6-8 are listed ligands (A) and corresponding rates for complexes undergoing the aquation

$$[Co(en)_2ACl]^+ + H_2O \rightarrow [Co(en)_2AOH_2]^{++} + Cl^-$$

The ligands are listed in order of conjugative ability, beginning

Table 6-8 *Rates of chloride release[a] by complexes of the type* $[Co(en)_2ACl]^+$

Ligand A		$k \times 10^7$, sec^{-1}
cis	OH$^-$	130,000
	Cl$^-$	1250
	NCS$^-$	114
	NH$_3$	4.5
	OH$_2$	150
	NO$_2^-$	1120
trans	OH$^-$	14,000
	Cl$^-$	160
	NCS$^-$	0.42
	NH$_3$	3.3
	NO$_2^-$	8900

[a] Data from Refs. 9 and 10.

with the electron-supplying OH$^-$; their order was deduced from known rates of organic nucleophilic substitution reactions. In both the *cis* series and the *trans* series, the rates pass through a minimum. The faster rates at the OH$^-$ end are ascribed to a predominantly S_N1 (bond-breaking) mechanism, where electron release stabilizes the five-coordinate intermediate through double-bonding as in (I). The rates decrease as the electron-releasing

(I)

power decreases since there is less stabilization of the transition state by double-bonding. On the other hand, as electron withdrawal becomes significant—with NO$_2^-$, for example—the *d* electrons of cobalt are withdrawn from the reaction site; nucleophilic attack by the water molecule becomes easier, and the rates increase again.

Only an introduction to bond breaking, bond formation, and

the nature of bonding in the transition state has been given here; further details should be obtained from recent reviews on the subject.[1-3, 9, 10]

6-5 BASE HYDROLYSIS

The problem of dissociation vs. displacement also comes up in base hydrolysis and in basic catalysis of replacement. Certain coordination compounds such as the cobaltammines, the aquo-chromium complexes (see Section 6-6), and the aquorhodium complexes undergo replacements with rates that are inversely proportional to hydrogen ion concentration even in acidic solutions of about pH 3. The actual action of hydroxide ion has been the subject of some debate since two types of mechanism have been postulated.

On the basis of the known nucleophilicity of hydroxide ion in displacements on carbon and on the basis of its negative charge and small size, one group of workers (whose views are summarized by Ingold[9]) suggest that the reaction is of the S_N2 type. The reaction

$$OH^- + [Co(en)_2NH_3Cl]^{++} \rightarrow [Co(en)_2NH_3OH]^{++} + Cl^-$$

is first-order each in hydroxide ion and complex ion concentrations. When the reactant complex has the *trans* configuration, 76 per cent of the product is *cis*; when the reactant has the optically active *cis* configuration, the product is 16 per cent *trans*, 24 per cent optically active *cis*, and the balance racemic *cis*.

These data have been interpreted in terms of an *edge-displacement* mechanism, which is schematically represented by the process

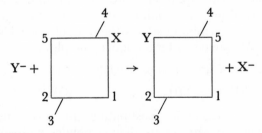

where Y^- is the entering ligand and X^- is the outgoing ligand. Depending on the placement of NH_3, several stereochemical isomers can be produced as observed. If NH_3 is at position 1 (*cis* complex), the product has the *trans* configuration. If NH_3 is at position 2, the *cis* configuration is obtained in the product. A *cis* configuration also results when NH_3 is at position 5.

This cannot be the sole mechanism since some *trans* product is formed from *trans* reactant, a result impossible by edge displacement. The *trans* product, however, can be formed by a front-side S_N2 attack of the type (II). As in all bimolecular substitutions, the

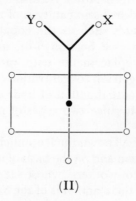

(II)

law of microscopic reversibility demands that for an exchange process the entering and leaving groups be symmetrically disposed.

The alternative mechanism for hydroxide ion reactions has three steps. The first is an equilibrium deprotonation of a ligand that is not the leaving group.

$$[Co(en)_2NH_3Cl]^{++} + OH^- \rightleftharpoons [Co(en)_2NH_2Cl]^+ + H_2O$$

Then follows the rapid loss of the chloride ion from the intermediate complex and finally aquation. This mechanism has been termed S_N1CB (substitution, nucleophilic, unimolecular, conjugate base) since the rate-determining step is the unimolecular loss of ligand from the conjugate base of the original complex.

$$[Co(en)_2NH_2Cl]^+ \rightarrow [Co(en)_2NH_2]^{++} + Cl^-$$

$$[Co(en)_2NH_2]^{++} + OH_2 \rightarrow [Co(en)_2NH_3OH]^{++}$$

The main reason for postulating this mechanism is that hydroxide ion is unique among all nucleophiles in showing second-order kinetics; if the mechanism were really S_N2, then contrary to usual observation one would expect other strong nucleophiles such as N_3^- and NCS^- to show replacement with second-order kinetics also. The evidence for the S_N1CB mechanism, which was initially proposed by Garrick[11] and extended by Basolo and Pearson,[1] is fairly convincing in some instances. As was seen in Chapter 4, hydroxide ion is a strong nucleophile but not uniquely so. In the base-catalyzed reactions of cobaltammines, however, it is unique, for it reacts at exceedingly low concentrations ($10^{-10} M$) whereas other nucleophiles are essentially nonreactive even at usual concentration levels.

Since a protonic hydrogen on a nonreactive ligand in the complex is a necessary part of the S_N1CB mechanism, cobalt complexes without protonic hydrogens should be particularly nonreactive in base hydrolysis, as indeed is the case. The rates of release of nitrite ion by the complex trans-dinitrobis (2,2′-bipyridine)cobalt(III) and of chloride ion by the complex trans-dichlorobis(tetraethylethylenediphosphine)cobalt(III) are independent of hydroxide ion concentration even at pH values of 11 to 12. Unfortunately this type of evidence is marred by the fact that some complexes with protonic hydrogens, which could react by an S_N1CB mechanism, show no reactivity to hydroxide ion; for example, the rate of hydrolysis of $[Cr(NH_3)_6]^{3+}$ is only twice as fast in 0.1 M NaOH as in 0.1 M HNO$_3$.

Another necessary part of the S_N1CB mechanism is a proton exchange more rapid than the base hydrolysis, because the first step is an equilibrium acid-base reaction. It has been found that the rates of hydrogen exchange are about 10^5 larger than the rates of chloride release in the complexes $[Co(NH_3)_5Cl]^{++}$ and cis-$[Co(en)_2NH_3Cl]^{++}$.

In any S_N1-type mechanism, the detection of an intermediate of lower coordination number is a possibility.† Although in water this is difficult because the solvent molecule is a good ligand and nucleophile, some nonaqueous, nonprotonic solvent might allow the detection. Recent studies[12] using dimethyl sulfoxide as solvent have given results that can be explained by an S_N1CB mechanism but not by an S_N2 process. It was observed that reactions of the type

$$[Co(en)_2NO_2Cl]^+ + Y^- \rightarrow [Co(en)_2NO_2Y]^+ + Cl^-$$

where Y^- is N_3^-, NO_2^-, or SCN^- are slow, with half-lives in hours; yet small amounts of hydroxide ion or piperidine catalyze the

† Isotope fractionation factors in the basic hydrolysis of some complex ions of the type $[Co(III)(NH_3)_5X]$ have been measured [M. Green and H. Taube, Inorg. Chem., **2**, 948 (1963)]. It was concluded that when X = Cl$^-$, Br$^-$, and NO$_3^-$, an S_N1CB mechanism occurs; when X = F$^-$, another type of mechanism obtains.

process and reduce the half-lives to minutes. Also, since the reaction of $[Co(en)_2NO_2OH]^+$ with Y^- is slow, a rapid S_N2 reaction of the initial complex with hydroxide ion to form a hydroxo complex is ruled out. Further, the reaction rate was found to be independent of the nature of Y^-. These results support the postulation of a five-coordinate intermediate, which is formed in a rate-determining step by loss of chloride ion from the conjugate base of the original complex, and which then in a rapid step picks up the ligand Y^-.

Even though evidence for the S_N1CB mechanism is strong, hydroxide ion does react with some complexes that have no acidic protons. For example, the complex ethylenediaminetetraacetato-cobalt(III) undergoes basic hydrolysis with second-order kinetics (first-order each in complex and hydroxide ion concentrations). This complex, which can be obtained in an optically active form, undergoes racemization by similar kinetics. These data are best explained by a bimolecular reaction in which a symmetrical seven-coordinate intermediate[13] is formed, with the hydroxide ion in the center of a rectangular face formed by the four acetate groups (III).

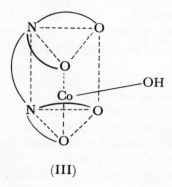

(III)

Furthermore, there are a few scattered cases where hydroxide ion and anions appear to act as true nucleophiles in displacements on transition-metal-complex cations. Margerum and Morgenthaler[14] have shown that tris(1,10-phenanthroline)iron(II) reacts with the three nucleophiles CN^-, OH^-, and N_3^- to give new complexes by loss of phenanthroline. The kinetics in each case are first-order in nucleophile and first-order in complex; the cyanide ion is most reactive and the azide ion least. An $S_N 1CB$ mechanism is not possible because there are no ionizable protons on the ligands; therefore the mechanism must involve a displacement. It is interesting to note that the authors[14] found for the analogous but not isoelectronic nickel complex a rate of dissociation independent of the nature or the concentration of the external ligand. This observation suggests that electronic structure is important in establishing the mechanism of replacement in coordination compounds.

6-6 CHROMIUM(III) COMPOUNDS

The rates of reaction for chromium(III) complexes have been carefully investigated. The rates are measurable, being about one power of ten faster than those of the analogous cobalt(III) complexes; for example, the aquations at 25°C of $[Co(NH_3)_5Br]^{++}$ and $[Cr(NH_3)_5Br]^{++}$ proceed with rate constants of 3.8×10^{-4} min^{-1} and 3×10^{-3} min^{-1}, respectively. This difference is to be expected because Co(III) with six electrons in the d orbitals has a higher CFSE in the ground state than does Cr(III) with three electrons. Another difference between the two cations is the greater ease with which chromiumammines hydrolyze to give $[Cr(OH_2)_6]^{3+}$.

The exchange of solvent water containing oxygen-18 with the aquochromium(III) ion proceeds slowly. The rate law

$$v = k[Cr(OH_2)_6^{3+}] \tag{6-2}$$

has been obtained,[15] although the rate constant is not independent of ionic strength as one might expect that of a unimolecular reaction to be. The change with salt concentration is not great, however; k is about 2×10^{-6} min^{-1} at 27°C and $\mu = 0.4$ and it increases to about 5×10^{-6} at $\mu = 9$. In the range of hydrogen ion concentration between 10^{-2} M and ~ 1.5 M, the rate is constant. The activation energy E_a is 24 ± 2 kcal mole^{-1}, and the pre-exponential factor is 1×10^{12} liter mole^{-1} sec^{-1}. It is not known whether this water exchange proceeds by an $S_N 2$ or by a dissociation-type mechanism. Either type fits the rate law, since the number of water molecules in the transition state is not indicated by the rate law.

The anation reaction

$$[Cr(OH_2)_6]^{3+} + NCS^- \rightleftharpoons [Cr(OH_2)_5NCS]^{++} + H_2O$$

has been carefully studied by Postmus and King.[16] This reaction, which proceeds slowly in aqueous solution from 14° to 95°C, has an equilibrium formation constant of about 1×10^3. The forward rate is

$$v_f = [Cr(OH_2)_6^{3+}][NCS^-](k_1 + k_2[H^+]^{-1} + k_3[H^+]^{-2}) \qquad (6\text{-}3)$$

and the rate law in the reverse direction is

$$v_r = [Cr(OH_2)_5NCS^{++}](k_{-1} + k_{-2}[H^+]^{-1} + k_{-3}[H^+]^{-2}) \qquad (6\text{-}4)$$

Therefore the activated complexes have the constitutions $[Cr(NCS)(OH_2)_x]^{++}$, $[Cr(NCS)(OH)(OH_2)_y]^+$, and $[Cr(NCS)(OH)_2(OH_2)_z]$. The values of the rate constants are such that the percentages of reaction carried by each of these three activated complexes are 14.4, 54.3, and 31.3, respectively, at $[H^+] = 1 \times 10^{-3}$ M. Since at this pH the predominant equilibrium species is $[Cr(OH_2)_6]^{3+}$, the activated complexes containing hydroxide ion are increasingly effective as compared to the trivalent ion itself.

Two mechanisms are consistent with the rate law for this anation. The S_N2 mechanism with an activated complex having a coordination number of seven is possible, as is also an S_N1 reaction involving initial loss of water followed by a competition between the thiocyanate ion and water for the intermediate. The rate law agrees with the mechanism

$$[Cr(OH_2)_6]^{3+} \rightleftharpoons [Cr(OH_2)_5]^{3+} + H_2O$$

$$[Cr(OH_2)_5]^{3+} + NCS^- \rightarrow [Cr(OH_2)_5NCS]^{++}$$

provided the equilibrium established in the first step is not seriously disturbed by the second. Since at the highest concentration of thiocyanate ion employed anation occurs only 4 per cent as fast as water exchange, the first step, which is the presumed path for water exchange, is not seriously disturbed. With the data at hand, it is not possible to distinguish between the two mechanisms; the base catalysis is nonetheless evidence for a dissociative mechanism of the S_N1CB type.

Bidentate ligands generally have tighter bonding, link at two coordination positions, and form optically active octahedra. For these reasons, their mechanisms have been of interest to the inorganic chemist. Trioxalatochromate(III) ion, an example of a complex with bidentate ligands, can undergo aquation, racemization, exchange with uncomplexed oxalate ion, and oxygen atom exchange with solvent water.

The aquation reaction

$$[Cr(C_2O_4)_3]^{3-} + 2H_2O \rightleftharpoons [Cr(C_2O_4)_2(OH_2)_2]^- + C_2O_4^=$$

proceeds with a measurable rate near 50°C. However, the equilibrium is almost completely to the left (K is 3.4×10^{-6} in the forward direction at 32°C), so that little product builds up. The rate of aquation has been investigated in acid solution,[17] where the equilibrium is shifted by protonation of product oxalate ion. The rate law is

$$v = [Cr(C_2O_4)_3^{3-}]\,(k_1[H^+] + k_2[H^+]^2) \qquad (6\text{-}5)$$

in the range of hydrogen ion concentration from $10^{-2}\ M$ to $0.97\ M$. The rate is independent of ionic strength and of added oxalic acid. The over-all activation energy is 22.1 kcal mole^{-1}. The data have been interpreted in terms of the mechanism

$$[Cr(C_2O_4)_3]^{3-} + H_3O^+ \rightleftharpoons [Cr(C_2O_4)_2 \cdot OC_2O_3H \cdot OH_2]^-$$

$$H_2O + [Cr(C_2O_4)_2 \cdot OC_2O_3H \cdot OH_2]^- \rightarrow$$
$$[Cr(C_2O_4)_2(OH_2)_2]^- + HC_2O_4^-$$

$$H_3O^+ + [Cr(C_2O_4)_2 \cdot OC_2O_3H \cdot OH_2]^- \rightarrow$$
$$[Cr(C_2O_4)_2(OH_2)_2]^- + H_2C_2O_4$$

with the first step postulated to be a rapid equilibrium; the second and third steps are competitive rate-determining steps. Evidence that the protonation step is an equilibrium is provided by the observation that this acid-catalyzed reaction is faster in D_2O than in H_2O. As mentioned earlier, D_3O^+ is a stronger acid than H_3O^+; therefore in D_2O a higher concentration of protonated intermediate builds up for reaction in the subsequent steps.

The reverse reaction (anation) is reported[18] to be independent of pH in the range $4.0 < pH < 9.3$, and also independent of oxalate ion concentration. Thus this reaction can be written in two steps

$$[Cr(C_2O_4)_2(OH_2)_2]^- + C_2O_4^= \rightleftharpoons [Cr(C_2O_4)_2 \cdot OC_2O_3 \cdot OH_2]^{3-} + H_2O$$

$$[Cr(C_2O_4)_2 \cdot OC_2O_3 \cdot OH_2]^{3-} \rightleftharpoons [Cr(C_2O_4)_3]^{3-} + H_2O$$

with the intermediate $[Cr(C_2O_4)_2 \cdot OC_2O_3 \cdot OH_2]^{3-}$ having one oxalate acting as a monodentate ligand. The rate-determining step in the formation of $[Cr(C_2O_4)_3]^{3-}$ is the closing of the chelate ring, with the formation of the intermediate from the diaquo complex being a faster step.

The same intermediate is again postulated in the exchange of oxalate ion with the tris complex. This process has been investigated by means of the carbon-14 isotope.[19] The data were found to fit a rate law of the form

$$v = [Cr(C_2O_4)_3^{3-}] \, (k_a + k_b[HC_2O_4^-] + k_c[H^+] + k_d[H^+][HC_2O_4^-])$$
(6-6)

for $2 < pH < 6$ and temperatures from $65°$ to $85°C$. A mechanism consistent with these results involves two rapid equilibria

$$[Cr(C_2O_4)_3]^{3-} + H_2O \rightleftharpoons [Cr(C_2O_4)_2 \cdot OC_2O_3 \cdot OH_2]^{3-}$$

$$[Cr(C_2O_4)_3]^{3-} + H_3O^+ \rightleftharpoons [Cr(C_2O_4)_2 \cdot OC_2O_3H \cdot OH_2]^-$$

and four rate steps

$$[Cr(C_2O_4)_2 \cdot OC_2O_3 \cdot OH_2]^{3-} + H_2O \rightleftharpoons$$
$$[Cr(C_2O_4)_2(OH_2)_2]^- + C_2O_4^-$$

$$[Cr(C_2O_4)_2 \cdot OC_2O_3 \cdot OH_2]^{3-} + HC_2^*O_4^- \rightleftharpoons$$
$$[Cr(C_2O_4)_2 \cdot OC_2^*O_3 \cdot OH_2]^{3-} + HC_2O_4^-$$

$$[Cr(C_2O_4)_2 \cdot OC_2O_3H \cdot OH_2]^- + H_2O \rightleftharpoons$$
$$[Cr(C_2O_4)_2(OH_2)_2]^- + HC_2O_4^-$$

$$[Cr(C_2O_4)_2 \cdot OC_2O_3H \cdot OH_2]^- + HC_2^*O_4^- \rightleftharpoons$$
$$[Cr(C_2O_4)_2 \cdot OC_2^*O_3H \cdot OH_2]^- + HC_2O_4^-$$

Assuming that all complexed species other than $[Cr(C_2O_4)_3]^{3-}$ are present in relatively small concentrations, it can be seen that each rate step corresponds to one term in the rate law. The third rate step here is identical to one postulated in the aquation reaction; the reported constants are identical, thus supporting the mechanistic postulations.

Racemization of the optically active tris complex $[Cr(C_2O_4)_3]^{3-}$ is known[20] to be more rapid than exchange, for no oxalate exchange occurs under conditions that give complete racemization. It is a necessary conclusion that racemization must occur by an intramolecular process which takes place without the oxalate ligands becoming completely free of the chromium at any time. There are several mechanisms by which an intramolecular racemization could occur, but the one that seems most probable for the present case is the dissociation of one end of the bidentate link to form a five-coordinate intermediate such as (IV). This intermediate has a plane of symmetry and

(IV)

therefore would return directly or indirectly to the racemic form of the tris complex. This intermediate also could be a precursor to the species $[Cr(C_2O_4)_2 \cdot OC_2O_3 \cdot OH_2]^{3-}$ postulated in the study of the oxalate-exchange and aquation-anation reactions. Further evidence for monodentate oxalates as intermediates in reactions of chelated species is the observation that all twelve oxygens in the tris complex exchange with labeled solvent water at a rate much faster than that for oxalate ion exchange.[21] Even though no exchange of solvent oxygen occurs with free oxalate ion, the exchange with the complexed oxalate proceeds at a rate almost equal to the rate of racemization.†

The photocatalyzed racemization and oxygen-exchange processes have recently been studied.[22] The quantum yields are not large, being of the order of 0.078 for racemization and 0.034 for oxygen exchange. The ratio of quantum yields is 2.3, which corresponds closely to the ratio of 2.6 observed for thermal rates of

† A study of the kinetics of racemization of *cis*-dioxalatodiaquochromate(III) has been reported [G. L. Welch and R. E. Hamm, *Inorg. Chem.*, **2**, 295 (1963)]. The results obtained suggest the presence of singly bonded oxalate in an intermediate complex; this is in agreement with the studies mentioned above.

the same processes. Similar solvent-isotope effects are also observed. The results of the photochemical experiments have been interpreted in terms of light-catalyzed formation of the intermediate $[Cr(C_2O_4)_2 \cdot OC_2O_3 \cdot OH_2]^{3-}$, which then reacts in the same manner as was postulated for the thermal processes.

6-7 COMPLEXES OF DIVALENT CATIONS

The release of a ligand from a complex should become easier as the charge on the central metal decreases; this is particularly true where bond breaking is predominant, as seems to be the case for transition metal complexes. It is not surprising, therefore, to find that complexes of divalent cations undergo replacement reactions more rapidly than do those of trivalent cations. For example, the complex $[Fe(EDTA)]^-$ is slow to exchange with Fe^{3+} whereas $[Fe(EDTA)]^=$ exchanges with Fe^{++} at a rate too fast to measure.

These rapid rates for divalent cations tend to mask crystal field effects unless rates for cations with the same charge are compared. It is not appropriate to compare rates for the d^8 system of octahedral Ni(II) with those for the inert d^3 and d^6 systems of chromium(III) and cobalt(III), even though the CFSE for each of the three ions is large. The nickel cation must be compared with other divalent cations of the same period to determine whether the predicted slower rates resulting from CFSE loss in the transition state can be observed. Some data on the reaction

$$M^{++}_{(aq)} \cdot SO_4^- \underset{k_2}{\overset{k_1}{\rightleftharpoons}} MSO_{4(aq)}^0$$

in which the ion pair on the left combines to form a true complex, are shown in Table 6-9. Eigen[23] studied these rapid rates using relaxation spectroscopy. For the alkaline earth cations, the rates increase markedly with cation size. Among the transition metal cations, whose sizes are all similar, nickel reacts most slowly as would be expected.

Also in agreement with the fact that nickel ion has the slowest rates among the spin-free, divalent, first-row transition metal complexes are the data shown in Table 6-10 for the rates of dissociation of tripyridyl complexes.[24] The manganese, copper, and zinc complexes dissociate at a rate too fast to measure; for the other three metals, the rates follow the order Fe > Co > Ni.

One would predict on the basis of loss of CFSE in the transition

Table 6-9 *Rate constants for sulfation of divalent cations*[a]

Metal	k_1, sec^{-1}	k_2, sec^{-1}
Be^{++}	1×10^2	1.3×10^3
Mg^{++}	1×10^5	8×10^5
Ca^{++}	$\sim 10^7$	$\sim 10^8$
Cu^{++}	$\sim 10^7$	1×10^6
Ni^{++}	1×10^4	1×10^5
Co^{++}	2×10^5	2.5×10^6
Mn^{++}	3×10^6	2×10^7
Fe^{++}	$\sim 1 \times 10^6$	

[a] Data (at 25°C in H_2O) from Ref. 23.

Table 6-10 *Rate parameters and CFSE changes for dissociation of mono(tripyridyl) metal complexes*[a]

Complex ion	$k \times 10^3$, min^{-1}	E_a	ΔE, Dq
[Mn(tripy)]$^{++}$	fast		0
[Fe(tripy)]$^{++}$	398	18	0
[Co(tripy)]$^{++}$	6.3	20	0
[Ni(tripy)]$^{++}$	0.0016	24	2
[Cu(tripy)]$^{++}$	fast		0
[Zn(tripy)]$^{++}$	fast		0

[a] Data from Ref. 24. Rates are at pH $\simeq 7$ and 25°C.

state[1] that iron(II) in the spin-paired state should react even more slowly than nickel(II); this is an observed fact. The ferrocyanide system is known to be inert,[4] and [Fe(tripy)$_2$]$^{++}$ dissociated at a rate[24] only one-tenth as fast as [Ni(tripy)$_2$]$^{++}$.

References

1. F. Basolo and R. G. Pearson, *Mechanisms of Inorganic Reactions*, Wiley, New York, 1958.
2. D. R. Stranks, in J. Lewis and R. G. Wilkins (eds.), *Modern Coordination Chemistry*, Wiley-Interscience, New York, 1960, Chap. 2.

3. F. Basolo and R. G. Pearson, in H. J. Emeleus and A. G. Sharpe (eds.), *Advances in Inorganic Chemistry and Radiochemistry*, Academic, New York, 1961, Vol. 3, pp. 1–89.
4. H. Taube, *Chem. Rev.*, **50**, 69 (1952).
5. L. C. Pauling, *Nature of the Chemical Bond*, 3d ed., Cornell University Press, Ithaca, N.Y., 1960.
6. F. Basolo, W. R. Matoush, and R. G. Pearson, *J. Am. Chem. Soc.*, **78**, 4883 (1956).
7. F. A. Posey and H. Taube, *J. Am. Chem. Soc.*, **79**, 225 (1957).
8. H. R. Hunt and H. Taube, *J. Am. Chem. Soc.*, **80**, 2642 (1958).
9. C. K. Ingold, Kekulé Symposium on Theoretical Organic Chemistry, Chemical Society, London, 1958.
10. C. Ingold, R. S. Nyholm, and M. L. Tobe, *Nature*, **187**, 477 (1960).
11. F. J. Garrick, *Nature*, **139**, 507 (1937).
12. R. G. Pearson, H. H. Schmidtke, and F. Basolo, *J. Am. Chem. Soc.*, **82**, 4434 (1960).
13. D. W. Cooke, Y. A. Im, and D. H. Busch, *Inorg. Chem.*, **1**, 13 (1962).
14. D. W. Margerum and L. P. Morgenthaler, in S. Kirschner (ed.), *Advances in the Chemistry of the Coordination Compounds*, Macmillan, New York, 1961, p. 481.
15. J. P. Hunt and R. A. Plane, *J. Am. Chem. Soc.*, **76**, 5960 (1954); J. P. Hunt and H. Taube, *J. Chem. Phys.*, **19**, 602 (1951); R. A. Plane and H. Taube, *J. Phys. Chem.*, **56**, 33 (1952).
16. C. Postmus and E. L. King, *J. Phys. Chem.*, **59**, 1208, 1217 (1955).
17. K. V. Krishnamurty and G. M. Harris, *J. Phys. Chem.*, **64**, 346 (1960).
18. R. E. Hamm and R. H. Perkins, *J. Am. Chem. Soc.*, **77**, 2083 (1955).
19. F. D. Graziano and G. M. Harris, *J. Phys. Chem.*, **63**, 330 (1959).
20. C. H. Johnson, *Trans. Faraday Soc.*, **31**, 1612 (1935); E. Bushra and C. H. Johnson, *J. Chem. Soc.*, **1939**, 1927; G. K. Schweitzer and J. L. Rose, Jr., *J. Phys. Chem.*, **56**, 428 (1952).
21. D. R. Llewellyn and A. L. Odell, *Australian At. Energy Symp.*, *Proc. Sydney*, **5**, 623 (1958).
22. S. T. Spees and A. W. Adamson, *Inorg. Chem.*, **1**, 531 (1962).
23. M. Eigen, *Z. Elektrochem.*, **64**, 115 (1960); see also M. Eigen, in S. Kirschner (ed.), *Advances in the Chemistry of the Coordination Compounds*, Macmillan, New York, 1961, p. 371.
24. R. Hogg, G. A. Melson, and R. G. Wilkins, in S. Kirschner (ed.), *Advances in the Chemistry of the Coordination Compounds*, Macmillan, New York, 1961, p. 391.

7

Electron-Transfer Reactions
of Complexes

Oxidation-reduction reactions and electron-exchange reactions of coordination compounds often involve replacements in the coordination sphere. This chapter is meant to serve as an introduction to these electron-transfer reactions; more detailed information is available in several recent reviews.[1-3, 4, 5]

7-1 ELECTRON-TRANSFER THEORY

In the gas phase, electron transfer between a rare-gas atom and its ion

$$A + A^+ \rightarrow A^+ + A$$

is efficient; experiments have shown that the effective diameter for electron transfer is larger than the collision diameter. The ease of electron transfer results from the extension into space of the orbitals and, except for the transferred electron, from the identity of the two particles. The rate of transfer increases as the amount of orbital overlap increases.

In the liquid phase, the situation is more complicated. Solvent molecules around the exchanging particles hinder the extension into space of the orbitals; ligands attached to ions are particularly effective in this shielding. In addition, the ligands are bound to the exchanging particles in a non-identical manner. For example, in the ground state for the electron exchange

$$[Fe(OH_2)_6]^{++} + [Fe^*(OH_2)_6]^{3+} \rightleftharpoons [Fe(OH_2)_6]^{3+} + [Fe^*(OH_2)_6]^{++}$$

the iron-oxygen bond distances in the Fe(III) ion are shorter than those in the Fe(II) ion. The Franck-Condon principle states that nuclear motion is slow as compared to electronic motion; therefore, in the normal state of these complexes, the transfer of an electron would result in an Fe(III) ion with elongated bonds and an Fe(II) ion with shortened bonds. Since the products of electron exchange would then have a higher energy than the reactants, the probability of the exchange is very small.

Exchange can occur, however, when the two particles have nearly identical structural and electronic configurations. If their configurations are not normally identical, the addition of energy through thermal motion can make the partners identical with respect to everything except the exchanging electron. The activation energies observed for these exchange processes in solution probably result from the bond distortion necessary to attain identity [squeezing in the water ligands around Fe(II) ion and extending the iron-oxygen bonds in aqueous Fe(III) ion] and from electrostatic repulsion between two ions of similar charge.

Except for the transferring electron, it is also important that the two ions have identical spin states. If with the electron exchange there is an over-all change in the spin, the transition is quantum-mechanically forbidden and can proceed only very slowly. For example, as expected, exchange is slow between ammine complexes of Co(II) and Co(III) because the Co(III) is in a spin-paired state whereas Co(II) has three unpaired electrons. For transfer to take place, one or both of the ions must be promoted to an excited state. This promotion, however, is difficult with ligands of intermediate field-strength such as NH_3, ethylenediamine, and oxalate ion. The electron-exchange rate constants at 25°C for the octahedral Co(II) and Co(III) complexes of these ligands are $< 10^{-8}$, 5×10^{-5}, and 9×10^{-7} liter mole^{-1} sec^{-1}, respectively. On the other hand, Co(III) complexes with ligands of weak field-strength such as water can be excited from a spin-paired to a spin-free state by a small energy input. Consequently, the electron exchange between the aquo complexes of cobalt is fast, with a rate constant of 0.75 liter mole^{-1} sec^{-1} at 0°C. In addition, Co(II) complexes with ligands of strong field-strength such as phenanthroline can be easily excited from the spin-free to the spin-paired state; the electron exchange for these cobalt complexes is again fast, with a rate constant of 1.1 liter mole^{-1} sec^{-1} at 0°C.

Marcus, Zwolinski, and Eyring[6] have derived a theory for electron-transfer reactions based on the hypothesis that the mechanism involves electron tunneling. Solvent and ligands produce an electronic energy barrier that the transferring electron must penetrate. In terms of the transition-state theory of chemical kinetics, their results may be written in the form

$$k = \frac{kT}{h} \, \kappa \exp \left(- \frac{\Delta F_r^{\ddagger}}{RT} - \frac{\Delta F_e^{\ddagger}}{RT} \right) \tag{7-1}$$

where κ is a transmission coefficient that includes the probability of barrier penetration, $\Delta F_r^{\ddagger}$ is the activation free-energy for rearrangement of the hydration and coordination shells, and $\Delta F_e^{\ddagger}$ is the activation free-energy for overcoming electric repulsion between the ions. The transmission coefficient, which always remains less than 1, increases (tending to increase the rate constant) as the exchanging partners come closer together. Because of electrostatic repulsion, the energy of activation also increases; however, this tends to decrease the rate. The net result is that at an optimum distance of about 6 A the exchange rate is at a maximum. Similar theories for electron transfer by a tunneling mechanism have been proposed by Weiss[7] and by Marcus.[8]

Ligands that can act as bridging groups facilitate electron transfer. They bind the two metal ions in an activated complex where orbital extension into space is not necessary for electron exchange because the electron can pass through the ligand itself. Some bridging ligands are H_2O, OH^-, $O^=$, $O_2^=$, F^-, Cl^-, Br^-, I^-, N_3^-, SCN^-, and the carboxylate ions. The ligand-bridge mechanism is not effective, however, unless ligand replacement in at least one of the ions is faster than the rate of electron transfer.

In some instances ligands can also act as electron-conduction media. Recent work on electron-spin resonance of paramagnetic complexes has shown that unpaired electrons spend part of their time on ligands. Consequently, electron transfer can occur by a mechanism involving movement of an electron from the outside of a ligand in one coordination sphere over to the outside of a second sphere. This mechanism seems particularly appropriate with large conjugated ligands like phenanthroline and bipyridine.

7-2   OUTER-SPHERE EXCHANGE REACTIONS

Exchange reactions in which both partners are inert to substitution but which show rapid rates of electron exchange must involve a mechanism related to electron tunneling. Taube[3] denotes this type as an *outer-sphere activated complex*, in keeping with the high probability that the coordination spheres of the exchanging partners are not altered during or by the electron-transfer process. Some examples are $IrCl_6^{3-}$ and $IrCl_6^{=}$, $[Fe(CN)_6]^{4-}$ and $[Fe(CN)_6]^{3-}$, $[Mo(CN)_8]^{4-}$ and $[Mo(CN)_8]^{3-}$, $[Fe(Ph)_3]^{++}$ and $[Fe(Ph)_3]^{3+}$, and $[Os(dipy)_3]^{++}$ and $[Os(dipy)_3]^{3+}$.

An example of complexes where both partners are substitution-inert and where exchange is rapid is the pair MnO_4^- and $MnO_4^=$. Electron exchange between these two anions has been carefully studied by Sheppard and Wahl.[9] (It was known from previous work that oxygen exchange with water of these oxyanions is slow. Symons[10] showed that when MnO_4^- is reduced to $MnO_4^=$ by labeled water containing OH^-, the isotopic composition of MnO_4^- is unchanged.) Some data on the exchange rates are presented in Table 7-1. The kinetics are first-order each in the concentrations of the two manganese anions. The value of E_a is 10.5 kcal mole^{-1}, and $\Delta S^{\ddagger}$ is -9 cal mole^{-1} deg^{-1} at an NaOH concentration of 0.16 M. The presence of inert anions has no effect on the exchange

Table 7-1   *Rates of MnO_4^- and $MnO_4^=$ exchange*

Medium[a]	k, liter mole^{-1} sec^{-1}
0.16 M LiOH	700
0.16 M NaOH	710 $\pm$ 30
0.16 M KOH	800
0.16 M CsOH	2470
0.08 M NaOH and 0.08 M CsOH	1730
0.16 M NaOH and 10^{-3} M Co(NH$_3$)$_6$Cl$_3$	1860

[a]All reactions in water at 0°C; data from Ref. 9.

rate, whereas cations show large and specific effects. The rate is better correlated with the concentration of specific cation than with ionic strength; similar behavior has been observed for other systems by Olsen and Simonsen.[11]

The data on the manganese system indicate that somehow an electron is rapidly transferred from $MnO_4^=$ to MnO_4^-. Although it has not been proved, tunneling seems to be the most probable mechanism. Possibly the most interesting thing about the results is the very large rate observed with Cs^+. This outstanding enhancement is much too large to be a general ion-atmosphere effect; an external electron bridge between the two anions apparently is formed by this large, polarizable cation. The enhancement of exchange rate by the $[Co(NH_3)_6]^{3+}$ ion can be attributed to ion pairing, particularly to $MnO_4^=$. Ion pairs between highly charged ions of opposite sign are well established.

Most electron-exchange reactions between complex ions where both partners are substitution-inert have been observed to possess very high second-order rate constants. The electron-transfer constant for the system $[Fe(CN)_6]^{4-}$ and $[Fe(CN)_6]^{3-}$ is about 1×10^3 liter mole^{-1} sec^{-1} at 4°C[12]; those for $[Os(dipy)_3]^{++}$ and $[Os(dipy)_3]^{3+}$ and for $[Fe(Ph)_3]^{++}$ and $[Fe(Ph)_3]^{3+}$ are greater than 10^5 liter mole^{-1} sec^{-1} at 0°C.[4] It is a characteristic of these inert, spin-paired systems that the change in oxidation state with electron transfer only slightly alters the interatomic distances. Therefore the activation energy for ligand reorganization is small, and on the theoretical grounds previously mentioned the rates are expected to be rapid.

The exchange between cobaltamine complexes probably proceeds by the outer-sphere type of mechanism. Cobalt(III) complexes are inert, and coordinated ligands such as ammonia and ethylenediamine cannot form bridges even though Co(II) is labile. The observation that after electron transfer the product Co(III) has six amine groups suggests that the nitrogen ligands are already attached to Co(II) before electron transfer. The slow exchange rate between the cobalt ions is a consequence of the difference of about 0.5 A in the cobalt-nitrogen bond lengths for the two complexes and, as mentioned earlier, of the difference in spin states of the two ions. In contrast to other outer-sphere activated complexes, there is a large free-energy of structural and electronic reorganization.

If it is thermodynamically favored, electron exchange may be

rapid for pairs of chemically nonidentical and inert partners.†
George and Irvine[13] have found several reactions such as

$$[Fe(dipy)_3]^{++} + [Ru(dipy)_3]^{3+} \rightarrow [Fe(dipy)_3]^{3+} + [Ru(dipy)_3]^{++}$$

that have rate constants greater than 10^5 liter mole^{-1} sec^{-1}.

7-3 BRIDGE MECHANISM

Stable polynuclear complexes with ligand bridges are well
known in metal ion chemistry, and it is interesting to note that the
complexes containing two atoms of the same metal in different
oxidation states usually are very deeply colored, thus suggesting
electron interchange between the two atoms. It is hardly surprising
that electron transfer from one metal ion to another can take place
readily if one ligand can be bonded to both metals simultaneously.

The detection of an activated complex (there may also be an
intermediate) that has a bridging ligand depends on certain
conditions first clearly delineated by Taube. One of the reactant
complexes must be labile to substitution, and the other must be
inert. After electron transfer, the initially labile partner must be
inert to substitution; the other usually becomes labile. The reaction

$$Co(III) + Cr(II) \rightarrow Co(II) + Cr(III)$$

satisfies these conditions, for $Co(III)$ and $Cr(III)$ are inert whereas
$Cr(II)$ and $Co(II)$ are labile. The bridging ligand is brought into
the activated complex by the inert reactant, as, for example,
Cl^- by $[Co(NH_3)_5Cl]^{++}$. It ends up on the product side as
$[Cr(OH_2)_5Cl]^{++}$.

The reaction in acid solution is

$$[Co(NH_3)_5Cl]^{++} + [Cr(OH_2)_6]^{++} + 5H_3O^+ \rightarrow$$

$$[Co(OH_2)_6]^{++} + [Cr(OH_2)_5Cl]^{++} + 5NH_4^+$$

and proceeds rapidly with the rate law

$$v = k[Co(NH_3)_5Cl^{++}][Cr(OH_2)_6^{++}] \tag{7-2}$$

† The rates of electron-transfer reactions for a group of chemically similar
reactions would be expected to vary with the free-energy changes of the reactions
[R. A. Marcus, *J. Phys. Chem.*, **67**, 853 (1963)]. With Ce(IV) as the oxidant,
and substituted tris-(phenanthroline) complexes of Fe(II) ions as reductants,
an excellent linear-free-energy relation is observed [G. Dulz and N. Sutin,
Inorg. Chem., **2**, 917 (1963)].

The activated complex presumably has the structure (I) and the electron is said to be transferred from chromium to cobalt, with the chloride ion moving in the opposite direction. Alternatively, one

$$
\left[\text{H}_3\text{N} \underset{\text{H}_3\text{N}}{\overset{\text{H}_3\text{N}}{-}} \text{Co} \underset{\text{NH}_3}{\overset{\text{NH}_3}{\diagdown}} \text{Cl} \underset{\text{H}_2\text{O}}{\overset{\text{H}_2\text{O}}{-}} \text{Cr} \underset{\text{OH}_2}{\overset{\text{OH}_2}{\diagdown}} \text{OH}_2 \right]^{4+}
$$

(I)

can say that a chlorine atom is transferred. However, since electronic motion is very rapid as compared to nuclear motion, such alternative descriptions of electron transfer are probably without significance. Nevertheless, there is no doubt that an oxidation-reduction reaction has occurred by means of a bridged activated complex.

The important result of this experiment[14] is that $[\text{Cr}(\text{OH}_2)_5\text{Cl}]^{++}$ rather than $[\text{Cr}(\text{OH}_2)_6]^{3+}$ is formed. Once the chromium is oxidized, further alterations in the coordination sphere take place slowly. Therefore the chlorine-chromium bond must be formed at or before the time the oxidation occurs. When the above reaction was repeated with radioactive chloride ion present in solution, practically none of the radioactivity appeared in the product, thus confirming the direct nature of the chlorine transfer.

For the complexes $[\text{Co}(\text{NH}_3)_5\text{L}]^{3+}$, surveys of the ligands L that act in the same capacity as Cl^- have been made.[15] Transfer to chromium is observed with F^-, Br^-, I^-, $\text{SO}_4^=$, N_3^-, NCS^-, carboxylate ions, PO_4^{3-}, $\text{P}_2\text{O}_7^{4-}$, OH^-, and probably also NO_3^- and H_2O. Since ammonia, when bound as a ligand in a coordination sphere, cannot form a bridge, it is expected that the hexammine Co(III) ion will not oxidize Cr(II) ion rapidly. Such is the case, for the rate of oxidation by the hexammine complex is over one-thousand-fold slower than that by the closely similar pentamminemonoaquo complex, which can form a bridge; the aquo complex is in turn slower than the hydroxo and chloro complexes, which form strong bridges.

The results obtained with carboxylato ligands are particularly significant, for they demonstrate electron transfer through conjugated systems.[3, 16] With simple ligands such as acetato, butyrato, benzoato, and even methyl succinato, the rates of oxidation of Cr(II) ion via the bridge mechanism are remarkably uniform, as may be seen in Table 3-1. The postulated structure for the

activated complex given in Chapter 3, page 42, is consistent with
the observation of carboxylato transfer from cobalt to chromium.
Each oxygen of the carboxylate group is bound to a metal ion,
since steric hindrance probably prevents a single oxygen from
serving as a bridge. When the ligand is fumarate or terephthalate,
the rate constant is larger than with a simple carboxylate; with
these two ligands, acid catalysis is observed. This is at first sight
surprising in view of the electrostatic repulsion between cations.
Apparently chromium attacks the carboxylate group at the end
remote from cobalt, and electron transfer takes place through the
conjugated system in an activated complex of the type (II). The

$$
\left[(H_3N)_5Co-O-C \begin{matrix} O \\ \diagdown \\ C=C \diagdown H \\ H \diagup \quad \diagdown C-O-Cr(OH_2)_5 \\ O \diagup \end{matrix} \right]^{3+}
$$

(II)

proton is believed to improve conjugation through the pi system by
adding to the carbonyl group adjacent to the cobalt. Further
evidence for this mechanism is the reaction between Cr(II) and
the methyl fumarato complex (III) of Co(III). The reaction

$$
\left[\begin{matrix} H_3N \diagdown \quad \diagup NH_3 \\ H_3N-Co-O-C \diagup O \\ H_3N \diagup \quad \diagdown NH_3 \quad \diagdown \\ \quad | \\ H \diagdown C=C \diagup H \\ \quad \\ O \diagdown C \diagdown O-CH_3 \end{matrix} \right]^{++}
$$

(III)

proceeds with hydrolysis of the ester linkage plus incorporation of
the methyl alcohol into the coordination sphere of the Cr(III)
product.

The other carefully studied reaction that proceeds by a
bridge mechanism involves electron exchange between various
Cr(III) species and the aquo Cr(II) ion. This reaction

$$[Cr(OH_2)_5L]^{3+} + [Cr^*(OH_2)_6]^{++} \rightarrow$$

$$[Cr(OH_2)_6]^{++} + [Cr^*(OH_2)_5L]^{3+}$$

can be followed by radioactive-chromium tracer. The assumption
that there are six equivalent water molecules in the coordination
sphere of Cr(II) ion may be in error, for a spin-free d^4 system could

also have tetragonal symmetry (four water molecules close to chromium in a plane and two water molecules further away on an axis perpendicular to the plane). The reaction of $[Cr(NH_3)_5L]^{3+}$ with $Cr(II)$ is an electron-transfer reaction, analogous to the reaction of $[Co(NH_3)_5L]^{3+}$ with $Cr(II)$ in that the ligand L is transferred and the five ammine groups are set free in the course of the reaction. Data for second-order rate constants of electron exchange in chromium systems are presented in Table 7-2.

Since in these systems the product of electron exchange is $[Cr(OH_2)_5L]^{3+}$, the bridge mechanism must be operative. This is also indicated by the relative rates given in the table. The ease of electron transfer follows the order $I^- > Br^- > Cl^- > F^-$. The differences in magnitude among the rates of halogen transfer is quite striking but reasonable, for the most polarizable ligand is expected to transfer most easily. The results with the two tri-atomic ligands are equally significant. The large rate for azide transfer stems from the symmetry of the ligand, the mobile electronic structure of the ligand, and the comparatively small repulsion between the two cations in the configuration (IV). Electron

$$\diagdown\!\!\!\diagup Cr\!-\!N\!-\!N\!-\!N\!-\!Cr\diagup\!\!\!\diagdown$$

(IV)

Table 7-2 *Rates of reaction of chromium(II) ion with some chromium(III) complexes*

Chromium(III) complex	Temp., °C	k, liter mole^{-1} sec^{-1}	Note
$[Cr(OH_2)_6]^{3+}$	24.5	$\leqslant 2 \times 10^{-5}$	a
$[Cr(OH_2)_5OH]^{++}$	24.5	0.7	a
$[Cr(OH_2)_5F]^{++}$	27	2.6×10^{-2}	b
$[Cr(OH_2)_5Cl]^{++}$	0	9.1	b
$[Cr(OH_2)_5Br]^{++}$	0	>60	b
$[Cr(OH_2)_5NCS]^{++}$	27	1.8×10^{-4}	b
$[Cr(OH_2)_5N_3]^{++}$	0	>1.2	b
$[Cr(NH_3)_5F]^{++}$	25	2.7×10^{-4}	c
$[Cr(NH_3)_5Cl]^{++}$	25	5.1×10^{-2}	c
$[Cr(NH_3)_5Br]^{++}$	25	3.2×10^{-1}	c
$[Cr(NH_3)_5I]^{++}$	25	5.5	c

[a] A. Anderson and N. A. Bonner, *J. Am. Chem. Soc.*, **76**, 3826 (1954).
[b] D. L. Ball and E. L. King, *J. Am. Chem. Soc.*, **80**, 1091 (1958).
[c] A. E. Ogard and H. Taube, *J. Am. Chem. Soc.*, **80**, 1084 (1958).

transfer through thiocyanate is not as fast, for the product of the rate-determining step has the weak chromium-sulfur bond, which probably rearranges quickly to the normal chromium-nitrogen bond.

Evidence favoring the formation of a bridged intermediate before the electron-transfer step includes the low activation energies and the negative entropies of activation for these reactions (see, for example, the data in Table 3-1). It is difficult to visualize how any reaction of the type

$$[Cr(OH_2)_6]^{++} \rightleftharpoons [Cr(OH_2)_5]^{++} + OH_2$$

$$[Cr(OH_2)_5]^{++} + [Co(NH_3)_5L]^{3+} \rightarrow products$$

could have a total activation energy as low as 3 or 4 kcal mole^{-1}, for the observed E_a must equal the sum of the energies of the two steps. Even if the Cr(II) ion has the tetragonal structure, the loss of one water molecule from the coordination shell would certainly require at least 10 kcal mole^{-1}. Therefore this mechanism is not compatible with the observed energy of activation.

The mechanism

$$[Co(NH_3)_5L]^{3+} + [Cr(OH_2)_6]^{++} \underset{k_2}{\overset{k_1}{\rightleftharpoons}}$$

$$[(H_3N)_5Co{-}L{-}Cr(OH_2)_5]^{5+} + H_2O$$

$$[(H_3N)_5Co{-}L{-}Cr(OH_2)_5]^{5+} + H_2O \overset{k_3}{\rightarrow}$$

$$[Co(NH_3)_5(OH_2)]^{++} + [Cr(OH_2)_5L]^{3+}$$

with its bridged intermediate is more compatible with a low activation energy, provided the formation of the intermediate is an exothermic step and $k_2 \gg k_3$, for then the overall $\Delta H^{\ddagger}$ equals ΔH_1 plus $\Delta H_{\frac{3}{2}}^{\ddagger}$. The condition $k_2 \gg k_3$ depends on k_3 having a more negative entropy of activation; this is in agreement with the negative entropies observed for electron-transfer reactions.

In double-bridged compounds such as the aluminium chloride dimer and the ion (V), the polyhedra about two metal ions share a common edge. To discover whether this type of structure occurs in electron-transfer reactions, several experiments have been carried out. By means of oxygen isotopes, the reactions of Cr(II) ion with cis-diaquotetramminecobalt(III) ion or cis-diaquobis(ethylenediamine)cobalt(III) ion were shown to result in the transfer of a single oxygen atom.[17] Thus an activated

(V)

complex involving two bridging groups is unimportant as compared to the complex involving a single bridging group.† Similar conclusions resulted[18] from a study of the reaction of *cis*-difluorotetra-aquochromium(III) ion with Cr(II) ion. Since one fluoride ion is released to solution at each electron transfer, the activated complex must have a single fluorine bridge.

7-4   IRON(II)-IRON(III) EXCHANGE

Electron transfer from an aqueous Fe(II) ion to an aqueous Fe(III) ion was mentioned earlier along with some of the complications that arise in solvent systems. This particularly interesting reaction has often been studied, yet the mechanistic details are still uncertain.

The iron-oxygen bond distance in the $[Fe(OH_2)_6]^{3+}$ ion is 2.05 A, whereas in the related Fe(II) ion this distance is 2.21 A; although these data are from X-ray determinations in the solid state, they probably closely approximate the solution values. Both ions contain iron in the spin-free state; there are five unpaired electrons in the Fe(III) cation and four unpaired electrons in the Fe(II) cation. As expected from these spin-free configurations, both hexaaquo ions are labile. These data and the fact that the rate of electron exchange is measurable suggest several possible mechanisms for the exchange.

Rates for the exchange between Fe(II) and various Fe(III) complexes are presented in Table 7-3. The data show several striking features. As expected on electrostatic grounds, the lowest rate is found with the aquo Fe(III) cation. Addition of one halide

† The existence of a double-bridged transition state has been demonstrated in the exchange of *cis*-diazidochromium(III) ion with chromium(II) ion [R. Snellgrove and E. L. King, *J. Am. Chem. Soc.*, **84**, 4609 (1962)]. See also, A. Haim, *J. Am. Chem. Soc.*, **85**, 1016 (1963).

Table 7-3 *Rates of electron exchange between iron(II) cation and various iron(III) complexes*

Iron(III) complex	Rate, liter mole^{-1} sec^{-1} 0°C	E_a kcal mole^{-1}	$\Delta S^{\ddagger}$ cal mole^{-1} deg^{-1}	Notes
$Fe^{3+}(aq)$	0.87	9.9	-25	a
$FeOH^{++}$	1010	7.4	-18	a
FeF^{++}	9.7	9.1	-21	b
$FeCl^{++}$	9.7	8.8	-24	a
$FeBr^{++}$	4.9	8.5	-25	c
FeF_2^{+}	2.5	9.5	-22	b
$FeCl_2^{+}$	15	9.7	-20	a
$FeBr_2^{+}$	19	14	-3	c
$FeNCS^{++}$	12.1	7.9	-27	d
$Fe(NCS)_2^{+}$	2.0	8.6	-28	d
FeN_3^{++}	1800	13.7	$+6$	e
$FeC_2O_4^{+}$	700	9.1	-14	c

[a] J. Silverman and R. W. Dodson, *J. Phys. Chem.*, **56**, 846 (1952).
[b] J. Hudis and A. C. Wahl, *J. Am. Chem. Soc.*, **75**, 4153 (1953).
[c] R. A. Horne, *J. Phys. Chem.*, **64**, 1512 (1960).
[d] G. S. Laurence, *Trans. Faraday Soc.*, **53**, 1326 (1957).
[e] D. Bunn, F. S. Dainton, and S. Duckworth, *Trans. Faraday Soc.*, **55**, 1267 (1959).

ion to the activated complex raises the rate by a factor of about ten; this also is expected on electrostatic grounds, for the resulting smaller charge on the Fe(III) complex lowers the repulsion between cations in the activated complex. Surprisingly, fluoride, chloride, and bromide each give approximately the same rate, both in the monohalo complexes and in the dihalo complexes. If a bridge mechanism were operative, the order Br$^-$ > Cl$^-$ > F$^-$ would be expected, as in the data of Table 7-2. By way of contrast, the much greater rates with hydroxide, azide, and oxalate ion suggest that a bridge is formed. It is possible that a combination of the bridge and outer-sphere mechanisms is operative; in this case the observed rates would not conform exactly to values expected for either single mechanism.†

† In a well-designed set of experiments on a related series of compounds, the presence of both outer-sphere and inner-sphere mechanisms has been demon-

Another possible mechanism for electron transfer between aquo cations is hydrogen atom transfer between hydration shells. For example, the exchange between $[Fe(OH_2)_6]^{++}$ and $[Fe^*(OH_2)_5OH]^{++}$ could proceed by an activated complex (VI),

$$\left[-\overset{/}{\underset{\diagdown}{Fe}}-O-H\cdots O-\overset{/}{\underset{\diagdown}{Fe^*}}- \right]^{4+}$$
$$\qquad\quad \underset{H}{|} \qquad \underset{H}{|}$$

(VI)

in which the transferred hydrogen atom is symmetrically disposed. The fact that the rates for exchange between the aquo ions—and between Fe(II) ion and several complex Fe(III) ions—are about twice as fast in H_2O as in D_2O is consistent with considerable stretching of the O—H bonds in the transition state.[19] For a time, this was considered strong evidence for a hydrogen-atom-transfer mechanism. However, it has been found in more recent studies that exchange reactions having an outer-sphere mechanism or a bridge mechanism can also have deuterium-isotope effects of similar magnitude.[5]

The mechanism for exchange between the Fe ions is still open to debate. The existing data are consistent with any of the three mechanisms (bridge, outer-sphere, and hydrogen-transfer) that have been proposed for electron-exchange reactions.

7-5 TWO-ELECTRON TRANSFERS

The exchange reactions discussed to this point involve a single-electron transfer from one transition metal ion to another; because these metals have five degenerate d orbitals, oxidation states differ by units of ± 1. For post-transition elements, the oxidation states change by ± 2 because compounds with unpaired electrons are not common. It is to be expected, therefore, that two-electron transfers would be prevalent.

The exchange between Tl(I) and Tl(III) ions has been

---

strated [J. P. Candlin, J. Halpern, and S. Nakamura, *J. Am. Chem. Soc.*, **85**, 2517 (1963)]. The reduction of $[Co(NH_3)_5X]^{3+}$ by $Co(CN)_5^{3-}$ shows two different mechanisms. The inner-sphere mechanism, where X is the bridging ligand, shows rate constants that are markedly dependent on the nature of X. In contrast, the outer-sphere mechanism operates with a rate constant that shows only a relatively small dependence on the nature of X.

studied in media containing a variety of ligands. In aqueous perchlorate solution[20] the rate law is

$$v = k_1[Tl^+][Tl^{3+}] + k_2[Tl^+][TlOH^{++}] \qquad (7-3)$$

In nitrate solution[21] the same rate terms are found, along with the new term $k_3[Tl^+][TlNO_3^{++}]$. In these reactions and in those below, the ligand is assigned to the Tl(III) coordination sphere; however, the ligand's position in the activated complex is not known because both thallium partners are labile. In solutions containing chloride ion,[20] the rate first decreases and then increases as the concentration of ligand increases. Similar behavior is observed with cyanide ion as ligand.[22] Sulfate ion also has been found to have specific effects.[23] At the present time, it is difficult to postulate a detailed mechanism for these exchánge reactions. The position of the ligand in the activated complex and the cause of the unusual variation in rate with ligand concentration are two interesting problems yet to be solved.

When the ligand is bromide ion, the observed rate law[3] is

$$v = k_1[Tl^+][Tl^{3+}] + k_2[TlBr_2^+] + k_3[TlBr_3] + k_4[TlBr_2^-][TlBr_4^-] \qquad (7-4)$$

The first term is the simple electron exchange between the two ions. A proposed activated complex for the last term is (VII). It is

$$\begin{bmatrix} Br & & Br & & Br \\ & Tl & & Tl & \\ Br & & Br & & Br \end{bmatrix}^=$$

(VII)

interesting to note that here a double-bridge activated complex is suggested. The other two terms in the rate law presumably represent the oxidation-reduction equilibria

$$Tl^{3+} + 2Br^- \rightleftharpoons Tl^+ + Br_2$$

$$Tl^{3+} + 3Br^- \rightleftharpoons Tl^+ + Br_3^-$$

where the rate step in each case is the reaction from left to right.

7-6 NONCOMPLEMENTARY REACTIONS

Reactions in which the oxidant and reductant change oxidation state by the same number of electron equivalents are termed *complementary*. When the oxidant and reductant differ in their equivalence changes, the reactions are termed *non-*

complementary and generally proceed by more complicated mechanisms.

For example, the hypothetical reaction

$$2A^+ + B^{++} \rightarrow 2A^{++} + B$$

where A^+ is a one-equivalent reductant and B^{++} is a two-equivalent oxidant may have a variety of mechanisms[5]:

1. *One-step termolecular mechanism*

    $$2A^+ + B^{++} \rightarrow 2A^{++} + B$$

2. *Bimolecular mechanism: initial one-electron step*

    $$A^+ + B^{++} \rightarrow A^{++} + B^+$$

    $$A^+ + B^+ \rightarrow A^{++} + B$$

3. *Bimolecular mechanism: initial two-electron step*

    $$A^+ + B^{++} \rightarrow A^{3+} + B$$

    $$A^+ + A^{3+} \rightarrow 2A^{++}$$

4. *Bimolecular mechanism: initial disproportionation*

    $$2A^+ \rightarrow A + A^{++}$$

    $$A + B^{++} \rightarrow A^{++} + B$$

In mechanisms 2, 3, and 4 the initial step involves the formation of an unstable species and presumably is the slower step.

Although mechanism 1 has not yet been observed in reactions involving only metal ions, it has been observed in several instances involving the oxidation or reduction of metal ions by oxygen or hydrogen, respectively. For example, the rate law

$$v = k[Ag^+]^2[H_2] \tag{7-5}$$

has been found for the hydrogen reduction of silver ion.[24]

Since mechanisms 2 and 3 have identical rate laws for the first step, it is not possible to distinguish between them unless the mechanism has some unusual aspect such as inhibition by a product. An example is the reaction[25]

$$2Fe(II) + Tl(III) \rightarrow 2Fe(III) + Tl(I)$$

which has the rate law

$$v = k_{obs} \frac{[Fe(II)]^2[Tl(III)]}{[Fe(II)] + k'[Fe(III)]} \tag{7-6}$$

The proposed mechanism is

$$Fe(II) + Tl(III) \underset{k_2}{\overset{k_1}{\rightleftharpoons}} Fe(III) + Tl(II)$$

$$Fe(II) + Tl(II) \overset{k_3}{\rightarrow} Fe(III) + Tl(I)$$

where $k_{obs} = k_1 k_3 / k_2$ and $k' = k_2 / k_3$. The presence of the intermediate Tl(II) is strongly indicated by the fact that the rate is repressed by Fe(III); therefore the reaction must follow mechanism 2.

An example of mechanism 4 is the exchange between Ag(I) and Ag(II), for which the rate law[26] is

$$v = k[Ag(II)]^2 \qquad (7\text{-}7)$$

The disproportionation mechanism

$$2Ag(II) \underset{k_2}{\overset{k_1}{\rightleftharpoons}} Ag(I) + Ag(III)$$

with k_1 as the rate-determining step and k_2 as a fast follow-up, has been proposed.

Schematic mechanisms similar to the four presented can be derived for any set of equivalence changes. The oxidation-reduction reactions of chromium are particularly interesting because the two stable oxidation states differ by three equivalents and because there are three unstable oxidation states. In oxidations of Cr(III) or in reductions of Cr(VI), these unstable states are expected to be present as reactive intermediates. For example, the reaction

$$Cr(III) + 3Ce(IV) \rightarrow Cr(VI) + 3Ce(III)$$

in acidic aqueous sulfate media has been found[27] to follow the rate law

$$v = k[Ce(IV)]^2[Cr(III)][Ce(III)]^{-1} \qquad (7\text{-}8)$$

This rate constant shows an approximately inverse square dependence upon the concentration of bisulfate ion. A reasonable mechanism for this reaction of a one-equivalent oxidant and a three-equivalent reductant is

$$Ce(IV) + Cr(III) \rightleftharpoons Ce(III) + Cr(IV)$$

$$Ce(IV) + Cr(IV) \rightarrow Ce(III) + Cr(V)$$

$$Ce(IV) + Cr(V) \rightarrow Ce(III) + Cr(VI)$$

The first step is presumed to be a rapid equilibrium, the second step is rate-determining, and the last step is fast.

By way of comparison, the reaction

$$3Fe(II) + Cr(VI) \rightarrow 3Fe(III) + Cr(III)$$

probably proceeds with the complicated rate law[28]

$$v = k[Cr(VI)]^{1.7}[Fe(II)]^2[H^+]^2[Fe(III)]^{-1} \tag{7-9}$$

The unusual dependence on $[Cr(VI)]$ suggests that a dimer similar to $Cr_2O_7^=$ is present in the transition state; we shall, however, ignore this complexity and the fact that further work on the rate law is necessary, and we shall consider the mechanism

$$Cr(VI) + Fe(II) \rightleftharpoons Cr(V) + Fe(III)$$

$$Cr(V) + Fe(II) \rightarrow Cr(IV) + Fe(III)$$

$$Cr(IV) + Fe(II) \rightarrow Cr(III) + Fe(III)$$

The first step is assumed to be a rapid equilibrium, the second step is rate-determining, and the last step is fast. The striking thing about the oxidation of $Cr(III)$ by the one-equivalent oxidant $Ce(IV)$ and the reduction of $Cr(VI)$ by the one-equivalent reductant $Fe(II)$ is that in both reactions the slow step involves the change between $Cr(IV)$ and $Cr(V)$. There is undoubtedly a large activation free-energy barrier for electron transfer between these two oxidation states. The coordination number of $Cr(III)$ is six, for the predominant species in acidic aqueous solution is $[Cr(OH_2)_6]^{3+}$. On the other hand, the coordination number of $Cr(VI)$ is four; the predominant species are $HCrO_4^-$ and $Cr_2O_7^=$, in both of which a chromium atom is surrounded by an oxygen tetrahedron. As pointed out by Tong and King,[27] a coordination-number change must occur somewhere in the sequence

$$[Cr(OH_2)_6]^{3+} \xrightarrow{-e} Cr(IV) \xrightarrow{-e} Cr(V) \xrightarrow{-e} HCrO_4^-$$

The observation that the slow step in two reactions involves the $Cr(IV)$–$Cr(V)$ stage suggests that the actual transfer of the electron is accompanied by significant rearrangement such as would accompany a coordination-number change. It has been concluded that $Cr(IV)$ is $[Cr(OH_2)_6]^{4+}$ and that $Cr(V)$ has the form of an oxyacid such as H_3CrO_4.

The exchange reaction

$$Cr(III)^* + Cr(VI) \rightleftharpoons Cr(VI)^* + Cr(III)$$

is a complementary reaction, but it is not a simple exchange because the oxidation states differ by three (simultaneous transfer of three electrons is unlikely) and the coordination numbers differ by two (forcing considerable atomic rearrangement). The exchange, which is indeed slow,[29] proceeds by the rate law

$$v = [Cr(OH_2)_6^{3+}]^{4/3}[H_2CrO_4]^{2/3}(k[H^+]^{-2} + k')$$ (7-10)

A mechanism consistent with these fractional orders is

$$Cr(III) + 2Cr(VI) \rightleftharpoons 3Cr(V)$$

$$Cr(V) + Cr(III)^* \rightarrow Cr(III) + Cr(V)^*$$

$$Cr(V)^* + Cr(VI) \rightarrow Cr(VI)^* + Cr(V)$$

where the second step is rate-determining. Chromium(V) in the first step is in chemical equilibrium with $Cr(III)$ and $Cr(VI)$. In the last step, exchange between $Cr(V)$ and $Cr(VI)$ is relatively rapid.

There are several points of interest in this reaction. First, fractional orders like those above are usually found in hetero-geneous reactions, but they also show up occasionally in mechanisms involving a rapid equilibrium before the rate step. Second, some exchange takes place when the rapid equilibrium of the first step is set up, but the bulk of the exchange takes place in the slow second step of the mechanism. Despite its rapidity, the first step does not carry the exchange because of the very small amount of $Cr(V)$ present at equilibrium. Third, the final step involves rapid electron exchange between $Cr(V)$ and $Cr(VI)$, both of which are presumably tetrahedral; as mentioned earlier, exchange between the species $MnO_4^=$ and MnO_4^-, which are isoelectronic with the chromium species, is very rapid.

7-7 REPLACEMENT THROUGH REDOX MECHANISM

Rates of ligand replacement vary markedly with the oxidation state of the metal ion. In the case of both chromium and cobalt, the divalent complexes are labile whereas the trivalent complexes are inert. Small amounts of the lower-oxidation-state species some-times make available a mechanism for bringing into equilibrium a greater amount of the higher-oxidation-state complex.

An example is the rapid exchange of water with the aquo $Co(III)$ ion, which because of its diamagnetism is expected to be substitution-inert. The exchange is thought to be catalyzed by the aquo Co(II) ion, small amounts of which are always present in solutions of $[Co(OH_2)_6]^{3+}$. The postulated mechanism is

$$[Co(OH_2)_6]^{++} + 6H_2O^* \rightleftharpoons [Co(H_2O^*)_6]^{++} + 6H_2O$$

$$[Co(H_2O^*)_6]^{++} + [Co(OH_2)_6]^{3+} \rightleftharpoons [Co(H_2O^*)_6]^{3+} + [Co(H_2O)_6]^{++}$$

This mechanism agrees with the facts that electron exchange between the two aquocobalt ions is rapid ($k_2 = 0.77$ liter mole^{-1} sec^{-1} at 0°C) and that the aquo Co(II) ion exchanges its water ligands with solvent at a rate too fast to measure by usual techniques.

Another example is the catalysis of reactions of the inert Cr(III) ion by Cr(II) ion. For instance, the ion $[Cr(OH_2)_4Cl_2]^+$ is transformed to $[Cr(OH_2)_5Cl]^{++}$ through the mechanism

$$[Cr(OH_2)_4Cl_2]^+ + [Cr(OH_2)_6]^{++} + H_2O \rightarrow$$
$$[Cr(OH_2)_6]^{++} + [Cr(OH_2)_5Cl]^{++} + Cl^-$$

The Cr(III) complex forms a Cl bridge to the Cr(II) ion; after electron transfer the resultant Cr(III) ion retains the bridging chlorine. In this way Cr(II) ion is regenerated each time the step occurs, and a large amount of the dichloro complex is converted to the monochloro complex by a small amount of Cr(II) ion.

It is also possible to form substituted Cr(III) species from mixtures of the hexaaquo ion and a ligand if a small amount of Cr(II) ion is added. The ligand is carried into an electron-transfer activated complex by the Cr(II) ion. The mechanism is written

$$[Cr(OH_2)_6]^{++} + L \rightleftharpoons [Cr(OH_2)_5L]^{++} + H_2O$$

$$[Cr(OH_2)_6]^{3+} + [Cr(OH_2)_5L]^{++} \rightarrow [Cr(OH_2)_6]^{++} + [Cr(OH_2)_5L]^{3+}$$

Ligands introduced into Cr(III) complexes in this manner include pyrophosphate and sulfate.[30] It should be noted that L does not act as a bridge in this reaction.

Replacements in amine Pt(IV) complexes are generally very slow, but they can be catalyzed by the addition of Pt(II). The rate law for the exchange of radiochloride between the aquochloride ion and several chloroamine Pt(IV) complexes is

$$v = k[Pt(IV)][Pt(II)][Cl^-] \tag{7-11}$$

The postulated mechanism involves the exchange of two electrons through a bridge intermediate.[31]

$$[Pt(NH_3)_4]^{++} + Cl^{*-} \rightleftharpoons [Pt(NH_3)_4Cl^*]^+ \qquad \text{fast}$$

$$[Pt(NH_3)_4Cl_2]^{++} + [Pt(NH_3)_4Cl^*]^+ \rightleftharpoons$$
$$[Cl-Pt(NH_3)_4-Cl-Pt(NH_3)_4-Cl^*]^{3+}$$
$$\updownarrow$$
$$[Cl-Pt(NH_3)_4]^+ + [Pt(NH_3)_4Cl_2^*]^{++}$$

The rate of chloride exchange for *trans*-$[Pt(NH_3)_4Cl_2]^{++}$ is two thousand times faster than that for the *cis* isomer and about ten thousand times faster than that for $[Pt(NH_3)_5Cl]^{3+}$; these facts agree with the expected weakening in the activated complex of the bond *trans* to the bridge and with the known fact that the Pt—N bond is stronger than the Pt—Cl bond. The mechanism is further supported by the observations that platinum exchange occurs at the same rate and that *trans*-$[Pt(NH_3)_4Cl_2]^{++}$ is formed in the catalyzed exchange of chloride ion in a hydrochloric acid solution of $[Pt(NH_3)_5Cl]^{3+}$.

The replacement reaction

$$trans\text{-}[Pt(en)_2Cl_2]^{++} + NO_2^- \rightarrow trans\text{-}[Pt(en)_2ClNO_2]^{++} + Cl^-$$

was thought at first to proceed by a simple S_N2 process, but it has recently been found[32] to have a very complicated mechanism. At 50°C, an induction period of several hours is observed. However, the reaction can be catalyzed by the addition of $[Pt(en)_2]^{++}$, which eliminates the induction period and gives the rate law

$$v = k[Pt(II)][Pt(IV)][NO_2^-] \tag{7-12}$$

This rate law is similar to that for the chloride exchange mentioned above [Eq. (7-11)]. By comparing predicted rates with observed rates, the authors[32] have been able to support the following mechanism when Pt(II) is absent.

$$[Pt(en)_2Cl_2]^{++} + NO_2^- + H_2O \rightarrow [Pt(en)_2]^{++} + NO_3^- + 2Cl^- + 2H^+$$

$$[Pt(en)_2]^{++} + NO_2^- \rightleftharpoons [Pt(en)_2NO_2]^+$$

Further support for this mechanism is gained from four facts. First, only one nitrite ion is introduced into the product; this is consistent with the inability of nitrite ion to act as a bridge. Second, the rates for nitrite introduction into various chloroamine Pt(IV) complexes parallel the rates for chloride exchange with the same complexes. Third, reduction but no substitution by nitrite ion of the very hindered complex $[Pt(tetrameen)_2Cl_2]^+$, where tetrameen is tetramethylethylenediamine, is observed. Fourth, from thermodynamic measurements, nitrite ion is known to be sufficiently powerful to reduce Pt(IV) complexes.

Labilization of an inert system by an unstable oxidation state is known. The slow chlorine exchange between chloride ion and $AuCl_4^-$ in pure solutions is enhanced by the addition of small amounts of the one-electron reducing agents Fe(II) and V(IV).[33] Two-electron reducing agents such as Sn(II) and Sb(III) are ineffective as is also Au(I). It has been concluded that some complex of Au(II), formed from Au(III) by one-electron reducing agents, is the active catalyst for the exchange.

Somewhat similar results have been observed with the $PtCl_6^=$ and Cl^- exchange[34] and with the reactions

$$PtX_6^= + 6I^- \rightarrow PtI_6^= + 6X^-$$

where X^- is Cl^- or Br^-.[35] These rates are greatly increased by light and by ferrous ion or thiosulfate ion but are inhibited by small amounts of $IrCl_6^=$ ion. A complex of Pt(III), possibly in the form $PtCl_5^=$, is postulated to be the catalyst for these reactions. A mechanism that explains the observations is

$$PtCl_6^= + Fe\,(II) \rightarrow PtCl_5^= + Fe(III) + Cl^-$$

$$PtCl_5^= + 5Cl^{*-} \rightleftharpoons PtCl_5^{*=} + 5Cl^-$$

$$PtCl_5^{*=} + PtCl_6^= \rightarrow PtCl_5^*Cl^= + PtCl_5^=$$

The second and third steps are chain propagations where each molecule of Pt(III) causes exchange for many $PtCl_6^=$ ions. Also of interest is the exchange between $PtCl_4^=$ and Cl^-, where the Pt(II) compound is labilized by the same Pt(III) intermediate.

References

1. F. Basolo and R. G. Pearson, *Mechanisms of Inorganic Reactions*, Wiley, New York, 1958, p. 303ff.
2. B. J. Zwolinski, R. J. Marcus, and H. Eyring, *Chem. Rev.*, **55**, 157 (1955).
3. H. Taube, in H. J. Emeleus and A. G. Sharpe (eds.), *Advances in Inorganic Chemistry and Radiochemistry*, Academic Press, New York, 1959, Vol. 1.

4. D. R. Stranks, in J. Lewis and R. G. Wilkins (eds.), *Modern Coordination Chemistry*, Wiley-Interscience, New York, 1960, Section V, p. 148ff.
5. J. Halpern, *Quart. Rev. (London)*, **15**, 207 (1961).
6. R. J. Marcus, B. J. Zwolinski, and H. Eyring, *J. Phys. Chem.*, **58**, 432 (1954).
7. J. Weiss, *Proc. Roy. Soc. (London)*, **A222**, 128 (1954).
8. R. A. Marcus, *J. Chem. Phys.*, **24**, 966 (1956); **26**, 867 (1957).
9. J. C. Sheppard and A. C. Wahl, *J. Am. Chem. Soc.*, **79**, 1020 (1957).
10. M. C. R. Symons, *J. Chem. Soc.*, **1954**, 3676.
11. A. R. Olson and T. R. Simonson, *J. Chem. Phys.*, **17**, 1167 (1949).
12. A. C. Wahl and C. F. Deck, *J. Am. Chem. Soc.*, **76**, 4054 (1954).
13. P. George and D. H. Irvine, *J. Chem. Soc.*, **1954**, 587.
14. H. Taube, H. Myers, and R. L. Rich, *J. Am. Chem. Soc.*, **75**, 4118 (1953).
15. H. Taube and H. Myers, *J. Am. Chem. Soc.*, **76**, 2103 (1954); H. Taube, *J. Am. Chem. Soc.*, **77**, 4481 (1955); R. K. Murmann, H. Taube, and F. A. Posey, *J. Am. Chem. Soc.*, **79**, 262 (1957).
16. R. T. M. Fraser, in S. Kirschner (ed.), *Advances in the Chemistry of the Co-ordination Compounds*, Macmillan, New York, 1961, p. 287; R. T. M. Fraser, *J. Am. Chem. Soc.*, **83**, 4920 (1961); D. K. Sebera and H. Taube, *J. Am. Chem. Soc.*, **83**, 1785 (1961).
17. W. Kruse and H. Taube, *J. Am. Chem. Soc.*, **82**, 526 (1960).
18. Y.-T. Chia and E. L. King, *Discussions Faraday Soc.*, **1960** (No. 29), 109.
19. J. Hudis and R. W. Dodson, *J. Am. Chem. Soc.*, **78**, 911 (1956).
20. G. Harbottle and R. W. Dodson, *J. Am. Chem. Soc.*, **70**, 880 (1948); **73**, 2442 (1951).
21. R. J. Prestwood and A. C. Wahl, *J. Am. Chem. Soc.*, **70**, 880 (1948); **71**, 3137 (1949).
22. E. Penna-Franca and R. W. Dodson, *J. Am. Chem. Soc.*, **77**, 2651 (1955).
23. C. H. Brubaker and J. P. Michael, *J. Inorg. Nucl. Chem.*, **4**, 55 (1957).
24. J. Halpern, *J. Phys. Chem.*, **63**, 398 (1959).
25. K. G. Ashurst and W. C. E. Higginson, *J. Chem. Soc.*, **1953**, 3044.
26. B. M. Gordon and A. C. Wahl, *J. Am. Chem. Soc.*, **80**, 273 (1958).
27. J. Y.-P. Tong and E. L. King, *J. Am. Chem. Soc.*, **82**, 3805 (1960).
28. F. H. Westheimer, *Chem. Rev.*, **45**, 419 (1949).
29. C. Altman and E. L. King, *J. Am. Chem. Soc.*, **83**, 2825 (1961).
30. J. B. Hunt and J. E. Earley, *J. Am. Chem. Soc.*, **82**, 5312 (1960).
31. F. Basolo, A. F. Messing, P. H. Wilks, R. G. Wilkins, and R. G. Pearson, *J. Inorg. Nucl. Chem.*, **8**, 203 (1958); F. Basolo, M. L. Morris, and R. G. Pearson, *Discussions Faraday Soc.*, **1960** (No. 29), 80.
32. H. R. Ellison, F. Basolo, and R. G. Pearson, *J. Am. Chem. Soc.*, **83**, 3943 (1961).
33. R. L. Rich and H. Taube, *J. Phys. Chem.*, **58**, 6 (1954).
34. R. L. Rich and H. Taube, *J. Am. Chem. Soc.*, **76**, 2608 (1954).
35. A. J. Poe and M. S. Vaidya, *J. Chem. Soc.*, **1961**, 2891.

8

Reactions of Oxyanions

Each oxyanion contains a central atom in a positive oxidation state, surrounded by oxide ions or hydroxide ions. Common examples are sulfate $SO_4^=$, nitrate NO_3^-, chromate $CrO_4^=$, and borate $B(OH)_4^-$. Although structurally related to coordination compounds, the oxyanions have mechanisms that differ from those discussed thus far. General factors relevant to oxyanion mechanisms, replacement rates (including rates of oxygen exchange† with solvent water), and oxidation-reduction reactions will be covered in this chapter.

8-1 FACTORS AFFECTING RATES

A factor of prime importance to rates of oxyanion reactions is the acidity of the media, for rates increase strikingly with hydrogen ion concentration.[1, 2] In fact, orders of two or more in hydrogen ion concentration are characteristic. One way of looking at this hydrogen ion catalysis is that the protons labilize the oxide ions that are breaking loose from the central atom.

† There is a comprehensive review on " Isotopic Tracers in Inorganic Chemistry" by D. R. Stranks and R. G. Wilkins [*Chem. Rev.*, **57**, 743–866 (1957)].

In the case of the chlorate ion ClO_3^-, for example, the protons can aid a dissociative-type mechanism

$$ClO_3^- + H^+ \rightleftharpoons HOClO_2$$

$$HOClO_2 + H^+ \rightarrow ClO_2^+ + H_2O$$

$$ClO_2^+ + X^- \rightarrow XClO_2$$

by changing an oxide ion into a water molecule, which can be easily lost. Similarly, an S_N2-type mechanism can be aided by protons as in the steps

$$ClO_3^- + H^+ \rightleftharpoons HOClO_2$$

$$HOClO_2 + H^+ \rightleftharpoons H_2OClO_2^+$$

$$H_2OClO_2^+ + X^- \rightarrow XClO_2 + H_2O$$

Presumably the role of the proton is to weaken the strong bond between the negative oxygen atom and the positive central atom. Less certain but reasonable is the conclusion that the strong bond is broken before or during the transition state.

Although the intermediate ClO_2^+ has not been identified in reactions of chlorate ion, electron-deficient intermediates are known for many oxyanions. The electrophiles CO_2, NO^+, NO_2^+, SO_2, and SO_3 are formed in some reactions of $CO_3^=$, NO_2^-, NO_3^-, $SO_3^=$, and $SO_4^=$, respectively. Other electrophiles such as Cl^+ and SeO_2 have been postulated as kinetic intermediates. All of these electrophiles can, in theory, be formed from an oxyanion by a stripping of oxide ions from the central atom through the action of protons.

Rates of replacement in the coordination sphere of an oxyanion should decrease as the basicity of the leaving group in the coordination sphere increases. Acid catalysis is related because the basicity of water (double-protonated oxide ion) is far less than that of oxide ion itself. Further evidence is provided by the hydrolysis rates for aryl sulfate ions.[3] The hydrolysis reaction

which probably takes place with breaking of the sulfur-oxygen bond rather than of the carbon-oxygen bond, follows the rate law

$$\frac{d[HSO_4^-]}{dt} = k[C_6H_5OSO_3^-][H^+] \tag{8-1}$$

The rates of hydrolysis of some ring-substituted aryl sulfate ions, the pK_a values for the corresponding phenols (in 30 per cent ethanol), and the sigma values of Hammett are presented in Table 8-1. It can be concluded that electron availability plays a lesser (though still significant) role in the hydrolysis of aryl sulfate ions than it does in the ionization of phenols. As expected, the rate of hydrolysis increases as the basicity of the leaving group (phenolate ion) decreases.

If the mechanism of replacement is either S_N1—with the electron-deficient intermediate in equilibrium with oxyanion—or S_N2, the reaction rate should depend on the nucleophilic power of the entering group. On the other hand, if the formation of the electron-deficient intermediate is rate-determining, the rate is independent of nucleophile concentration. Both types of rate law are observed, and at least three mechanisms are needed to explain the observations.

In the case of carbonates, the presence of free carbon dioxide in equilibrium with carbonic acid, H_2CO_3, has been repeatedly

Table 8-1 *Effect of ring substituents on the basicity of phenols and on the hydrolysis rates of aryl sulfate ions*

Substituent	σ^a	$pK_a(XC_6H_4OH)^b$	$k^c \times 10^5$, liter mole^{-1} sec^{-1}
p-NO$_2$	1.27	7.69	44.5
m-NO$_2$	0.710	9.02	20.5
m-Cl	0.373	9.89	16.3
p-Br	0.232	10.15	11.5
p-Cl	0.227	10.20	10.8
m-OCH$_3$	0.115	10.67	10.2
H	0.000	10.85	8.9
m-CH$_3$	-0.069	10.98	8.8
p-CH$_3$	-0.170	11.14	6.87
p-OCH$_3$	-0.268		4.50

[a] L. P. Hammett, *Physical Organic Chemistry*, McGraw-Hill, New York, 1940, p. 190.
[b] R. A. Benkeser and H. R. Krysiak, *J. Am. Chem. Soc.*, **75**, 2423 (1953).
[c] Data (at 48.6°C) from Ref. 3.

demonstrated.[4] The rate step in oxygen atom exchange with water is the dehydration step

$$H_2CO_3 \rightarrow CO_2 + H_2O$$

and the product CO_2 reacts with nucleophiles at rates that depend somewhat on basicity. As seen in Table 8-2, the rate constant increases as the basicity of the incoming nucleophile increases, but nucleophilic order depends on other factors as well, for dimethylamine and piperidine are more reactive than the more basic hydroxide ion.

The reaction of chlorate ion with halide ions (in the presence of arsenious acid) follows the rate law

$$v = k[ClO_3^-][X^-][H^+]^2 \tag{8-2}$$

The oxygen-isotope exchange of chlorate ion with water also is dependent on the square of the hydrogen ion concentration; however, it is slower than halide ion oxidation.[5] The exchange rate in D_2O was found to be 2.83 times as fast as that in normal water; this suggests that protonation of chlorate ion occurs prior to the rate step. The mechanism

$$2H^+ + ClO_3^- \rightleftharpoons H_2ClO_3^+$$

$$X^- + H_2ClO_3^+ \rightarrow products$$

Table 8-2   *Basicity and rates of reaction of nucleophiles with* CO_2

Nucleophile	$pK_b(18°C)$	$k_2(18°C)$,[a] liter mole^{-1} sec^{-1}
H_2O	15.8	2×10^{-4}
$C_6H_5NH_2$	9.3	3
NH_3	4.8	6×10^2
$C_6H_5CH_2NH_2$	4.7	2×10^3
$CH_2{=}CHCH_2NH_2$	4.4	2×10^3
$H_2NCH_2CO_2^-$	4.2	1.1×10^3
CH_3NH_2	3.4	1.4×10^3
$(CH_3)_2NH$	3.2	1.1×10^4
$C_2H_{10}NH$	3.0	1.3×10^4
OH^-	-1.8	2×10^3

[a] Data summarized in Ref. 2.

where the first step is rapid and reversible and the second step is rate-determining, fits the experimental data. From the data, it is not possible to say whether the halide ion attacks oxygen to give HOX and $HClO_2$ or attacks chlorine to give $XClO_2$ and H_2O. It is certain, however, that the intermediate ClO_2^+ is not formed prior to the rate step in the halide ion oxidations, although it is still a possible intermediate in the oxygen exchange.

The third type of mechanism that has been demonstrated is the rate-determining formation of an electron-deficient intermediate. Under certain conditions, the nitration of aromatic compounds has been found to follow the rate law

$$v = k[HNO_3][H^+] \tag{8-3}$$

with no dependence on the concentration of the benzene derivative. The mechanism must be

$$HNO_3 + H^+ \rightarrow NO_2^+ + H_2O$$

$$NO_2^+ + ArH \rightarrow ArNO_2 + H^+$$

where ArH is the aromatic compound and the first step is rate-determining. Some reactions of nitrite ion also have been found to be acid-catalyzed and to be zero-order in substrate; presumably, NO^+ or some related electrophilic intermediate is formed in the rate step.

Another factor important to the rates of oxyanion reactions is the oxidation state of the central atom. In cases of elements that form oxyanions in more than one oxidation state, it is found that the rates for the lower oxidation state are faster than those for the higher oxidation state. Rates of replacement in substituted sulfates are almost always slow, whereas replacements in substituted sulfites are normally too fast to measure; the results with selenium, the next higher congener to sulfur, are similar. At the same acidity, reaction rates of nitrite ion are faster than those of nitrate ion. Finally, rates for reactions of the chlorine anions probably decrease in the order $ClO^- > ClO_2^- > ClO_3^- > ClO_4^-$. There is no doubt that the perchlorate reactions are extremely slow, that hypochlorite reactions are rapid, and that the other two have rates that are intermediate.

A factor related to the one above is the charge on the central atom in an isoelectronic series. In a particular period of the periodic table, it is a good generalization that rates of replacements decrease as the charge increases from left to right, as is shown by the series

$H_2SiO_4^=$, $HPO_4^=$, $SO_4^=$, and ClO_4^-. Silicate ion exchanges its oxide ions rapidly with water, and other replacement reactions of silicates appear to be fast. Phosphate replacements are slow but seem to be faster than analogous sulfate replacements. Substitutions in perchlorates are exceedingly slow; for example, it is estimated[5] that in 9 M perchloric acid at room temperature the oxygen exchange with water has a half-time of greater than 100 years. This correlation of rate with charge is reasonable, for the strength of the oxygen-to-central-atom bond should increase as the charge on the central atom increases.

In any family of the periodic table, the rates of replacement increase as the size of the central atom increases. This is to be expected, since the strength of the oxygen-to-central-atom bond decreases as size increases. The data for the halate ions exemplify this trend. Chlorate ions exchange oxygens with water in acid solution at a measurable rate at 100°C, bromate ion exchange is measurable in acid solution at 30°C, and iodate ion exchanges rapidly at room temperature even in neutral solution.

The several factors that influence rates of oxyanion reactions all indicate that an important part of the mechanism of replacement is breaking of the central-atom-to-oxygen bond. Here there is a similarity to ligand replacement in coordination compounds, where bond breaking is a predominant aspect of the mechanism. On the other hand, oxyanions differ from coordination compounds in that replacement rates for the former strongly increase with hydrogen ion concentration; this is, of course, a consequence of the strongly basic nature of the oxide ion.

It is worth noting that the same factors govern oxidation-reduction reactions of oxyanions as govern oxygen-exchange reactions. Apparently the redox reactions proceed by mechanisms in which the reductant (nucleophile) forms a bond to the central atom of the oxyanion in (or prior to) the rate step.

8-2 OXYGEN EXCHANGE

In this section, the rate laws for oxygen exchange between oxyanions and water will be listed. Some conclusions as to mechanism are possible and will be mentioned, but most of the mechanisms are tentative.

For oxyanions XO_m^{n-} that are derived from small, highly charged central atoms, the leading rate law for exchange is

$$v = k[XO_m^{n-}][H^+]^2 \tag{8-4}$$

Examples include $SO_4^=$,[6] NO_3^-,[7,8] ClO_3^-,[5] BrO_3^-,[8,9] $CO_3^=$,[10] and NO_2^-.[11] In view of the fact that both SO_3 and CO_2 are stable species, it seems probable in these two cases that the rate-determining steps are of the S_N1 type as in

$$H_2SO_4 \rightarrow SO_3 + H_2O \qquad \text{slow}$$

$$SO_3 + H_2O^* \rightarrow H_2SO_3O^* \qquad \text{fast}$$

with exchange being accomplished by the fast second step. For the three ions of the type XO_3^-, the electron-deficient intermediates should be less stable (although the ions NO^+ and NO_2^+ are known) and exchange could proceed by either an S_N1 or an S_N2 mechanism. A piece of evidence favoring the S_N2 mechanism is the observation that chloride ion catalyzes the oxygen exchange between water and the ions NO_3^- and BrO_3^-.[8] This catalysis, which proceeds by the rate law

$$v = k[XO_3^-][H^+]^2[Cl^-] \tag{8-5}$$

suggests the mechanism

$$2H^+ + XO_3^- \rightleftharpoons H_2XO_3^+$$

$$H_2XO_3^+ + Cl^- \rightarrow ClXO_2 + H_2O$$

$$ClXO_2 + H_2O^* \rightarrow H_2XO_3^{*+} + Cl^-$$

with the second step (which is an S_N2 displacement) being rate-determining. By analogy, the noncatalyzed exchange could proceed by the mechanism

$$H_2XO_3^+ + H_2O^* \rightarrow H_2XO_3^{*+} + H_2O$$

with a direct displacement of water by water.

The exchange of nitrite ion with water[11] is more complicated. This exchange and other reactions of nitrite ion with donor particles will be discussed in Chapter 10.

When the central atom is larger and/or less highly charged, the second-order dependence on hydrogen ion concentration is no longer characteristic.† The leading term in the rate law for exchange is now first-order in hydrogen ion concentration. This rate law

$$v = k[XO_m^-][H^+] \tag{8-6}$$

† The oxygen exchange between chromate and water goes by at least two paths [M. R. Baloga and J. E. Earley, *J. Phys. Chem.*, **67**, 964 (1963)]. One path involves $CrO_4^=$ and is independent of pH; the other path involves $Cr_2O_7^=$.

has been observed when XO_m^- is OCl^-,[12] OBr^-,[12] ReO_4^-,[13] and IO_3^-.[14] With hypochlorite ion, the rate term $k'[HOCl][Cl^-]$ is also found; this may represent the step

$$HOCl + Cl^- \rightleftharpoons OH^- + Cl_2$$

Similarly, the exchange of hypobromite with water is found to be catalyzed by chloride ion and bromide ion.

The oxygen exchange between iodate ion and water is subject to base catalysis as well as acid catalysis.[14] General acids (such as NH_4^+ and H_3BO_3) and general bases (such as acetate ion, pyridine, NH_3, and CN^-) all are active catalysts for the iodate exchange. The possibility that iodate species such as $H_2IO_4^-$ (with iodine having a coordination number greater than three) are intermediates has been suggested,[14] although these intermediates are probably present in comparatively small quantities.

It is of interest to compare with the oxygen-exchange rates the rates for halogen atom exchange between halide ion and hypohalite. The chlorine exchange

$$OCl^- + Cl^{*-} \rightleftharpoons OCl^{*-} + Cl^-$$

follows the rate law

$$v = k[HOCl][Cl^-][H^+] \tag{8-7}$$

whereas the analogous bromine exchange follows the law

$$v = k[HOBr][Br^-] \tag{8-8}$$

Again the smaller oxyanion demonstrates the higher dependence on hydrogen ion concentration for reaction. The bromine exchange can be accomplished either through a step of the type

$$Br^- + HOBr \rightleftharpoons OH^- + Br_2$$

involving bromide ion attack on the positive bromine, or through the activated complex (I), which involves a nucleophilic attack on oxygen. The fact that the rates of bromine exchange and oxygen

$$\left[\begin{array}{c} Br\text{---}O\text{---}Br \\ | \\ H \end{array} \right]^-$$

(I)

exchange are closely similar suggests that both exchanges proceed by the same mechanism and that bromine molecule is an intermediate. A discussion of related problems in hypochlorite reactions is found in a recent review by Taube.[15]

8-3   EXCHANGE BETWEEN PHOSPHATE AND WATER

The exchange reaction

$$H_3PO_4 + H_2O^* \rightleftharpoons H_3PO_4^* + H_2O$$

is slow even at 100°C and is catalyzed by acid. Several groups of chemists have studied the reaction qualitatively, and recently two groups have quantitatively investigated the rates. There are several mechanistic paths by which oxygen exchange can proceed.

In concentrated phosphoric acid solutions (6 M to 18 M H_3PO_4), the rate increases rapidly with H_3PO_4 concentration.[16] Addition of $H_2PO_4^-$ has no effect on the rate, and addition of $HClO_4$ has only a small rate-enhancing effect. It has been shown that, in concentrated phosphoric acid solutions, the rate of oxygen exchange and the rate of pyrophosphoric acid ($H_4P_2O_7$) hydrolysis are nearly equal. The kinetic data have been interpreted in terms of the rate law

$$v = k_1 a_1 a_w + k_2 a_1^2 + k_3 a_1^3 a_w^{-1} \tag{8-9}$$

where a_1 is the activity of phosphoric acid and a_w is the activity of water. The first term was attributed to a path involving direct exchange between phosphoric acid and water; the second term (which predominates in acid strengths between 13 M and 18 M) was attributed to the reversible reaction

$$2H_3PO_4 \rightleftharpoons H_4P_2O_7 + H_2O$$

and the third term was attributed to the reversible equilibrium

$$3H_3PO_4 \rightleftharpoons H_5P_3O_{10} + 2H_2O$$

To the extent that activity coefficients of the transition state were not taken into account, the rate law may be in error; it seems certain, however, that much of the exchange is accomplished via paths involving condensed phosphoric acids.

The other investigation[17] was carried out on solutions more dilute in phosphate in order to compare the exchange results with rates of hydrolysis of monoalkyl phosphates and other organic phosphates.[18] The rate of exchange in the pH range from 1 to 9 passes through a maximum. This maximum at about pH 5 is attributed to the species $H_2PO_4^-$, and it follows that this species exchanges more rapidly than does either H_3PO_4 or $HPO_4^=$. The rates of hydrolysis of monoalkyl esters of phosphoric acid also have a maximum at the pH where the mononegative anion is the predominant species, and the rate constant of this process is insensitive to the electronic effect of the alkyl group. To explain these facts requires a mechanism that involves two protons only (or in the case of the monoalkyl phosphates, an alkyl group and one proton) since either addition or subtraction of one proton decreases the rate. The mechanism

$$H_2PO_4^- \rightleftharpoons \overset{O}{\underset{O}{\overset{\|}{P}}}-O-H \rightarrow PO_3^- + H_2O$$
$$\qquad\qquad \underset{O-H}{}$$

$$PO_3^- + H_2O^* \rightarrow H_2PO_3O^{*-}$$

with the unknown metaphosphate anion as an intermediate formed in the rate step, gives a good account of the facts. The metaphosphate ion is mononegative and would be expected to form from a mononegative orthophosphate derivative. The familiar nitrate ion is stabilized in the mononegative form, and it seems reasonable that an analogous trigonal planar phosphate ion could exist as an intermediate. This intermediate should, however, be non-isolable because it would take up water rapidly.

Evidence was also found[17] for a slower rate of exchange involving H_3PO_4 and for an acid-catalyzed exchange. For the latter path, the rate law

$$v = k[H_3PO_4][H^+] \tag{8-10}$$

was proposed, although the constants obtained varied somewhat because of the high ionic strengths (1 M to 9 M $HClO_4$) involved. Two mechanisms of oxygen exchange for the latter two cases seem plausible: a synchronous displacement of a water molecule (from H_3PO_4 or $H_4PO_4^+$) by a water molecule from the solvent, or formation of a five-coordinate phosphorus intermediate. No choice between these alternatives seems possible on the basis of the available data.

8-4   INDUCED REACTIONS

Mohr, almost a century ago, showed that sodium arsenite in basic solution is not oxidized by air although sodium sulfite is rapidly oxidized under the same conditions. In a mixture of the two salts, however, the sodium arsenite is oxidized along with the sodium sulfite. This phenomenon in which a relatively fast reaction between two substances A and B "forces" an otherwise very slow reaction between A and C is called an *induced reaction*. The reactant A that appears in both reactions is called the *actor*. The *inductor* B is the reagent that readily reacts with the actor. The reagent C that reacts slowly (or not at all) with the actor in the absence of an inductor is called the *acceptor*. In the above example, oxygen, sulfite ion, and arsenite ion are actor, inductor, and acceptor, respectively.

The phenomenon of induced reactions offers one of the best methods for showing the presence of unstable intermediates in reaction mechanisms. The amount of acceptor consumed as compared with the amount of inductor consumed provides detailed information about the nature of the intermediate. In any induced oxidation, the *induction factor* is defined as the ratio of the number of equivalents of acceptor oxidized to the number of equivalents of inductor oxidized. At the limiting high concentration of acceptor, the induction factor asymptotically approaches a value that is a ratio of small whole numbers.

Some data for the induced reduction of oxygen using the chromate-arsenite couple[19] are given in Table 8-3. The data show how the induction factor I_f, defined as the ratio between the number of equivalents of acceptor (oxygen) reduced and the number of equivalents of inductor (chromate ion) reduced, varies with the ratio between the initial concentrations of these species. The limiting value of the induction factor is apparently $\frac{4}{3}$, which means that one oxygen molecule is reduced for each chromate ion reduced. A mechanism for this induced reaction will be presented in Section 8-5.

An induced reaction that provides an interesting visual demonstration is the induced oxidation of iodide ion by chromate ion, with ferrous ion as the inductor. In dilute acid the simple oxidations of iodide ion by chromate ion

$$2HCrO_4^- + 6I^- + 14H^+ \rightarrow 2Cr^{3+} + 3I_2 + 8H_2O$$

and by ferric ion

$$2Fe^{3+} + 2I^- \rightarrow 2Fe^{++} + I_2$$

Table 8-3 *Induced reduction of oxygen by the alkaline chromate ion-arsenic trioxide system*[a]

Oxygen pressure, atm	Initial chromate, mmoles liter^{-1}	Ratio, atm liter mmole^{-1}	I_f
0.2	5.60	0.036	0.27[b]
0.2	3.24	0.062	0.45
0.2	1.30	0.154	0.81
1.0	5.60	0.178	0.68[b]
1.0	2.33	0.43	0.89
1.0	1.62	0.62	1.02[b]
1.0	1.08	0.92	1.21
1.0	0.55	1.82	1.21[b]
1.0	0.333	3.0	1.04
10	1.62	6.2	1.04
15	2.33	6.4	1.22
15	1.62	9.3	1.33

[a] Data (in borax buffer at $\mu = 1.75$ and 40.0°C) from Ref. 19.
[b] Average value from several experiments.

are slow, whereas under the same conditions the oxidation of ferrous ion by chromate ion

$$HCrO_4^- + 3Fe^{++} + 7H^+ \rightarrow Cr^{3+} + 3Fe^{3+} + 4H_2O$$

is rapid. When the chromate, ferrous, and iodide ions are all mixed in dilute acid, a rapid formation of iodine occurs and this is vividly demonstrated by the sudden appearance of the deep brown color of triiodide ion in aqueous solution. The induction factor approaches a value of 2, which indicates the stoichiometry

$$HCrO_4^- + Fe^{++} + 2I^- + 7H^+ \rightarrow Cr^{3+} + Fe^{3+} + I_2 + 4H_2O$$

for the rapid over-all reaction.

Of the three equivalents of oxidizing power for each chromate ion, only one is used by ferrous ion; the other two are used in oxidizing iodide ion. It seems certain from this fact and from the fact that iodide oxidation by chromate alone is negligible under these same conditions that a pentavalent chromium intermediate is present in the reaction sequence. The mechanism proposed by Westheimer[20] in his excellent review of oxidation mechanisms involving chromates is

$$Cr(VI) + Fe(II) \rightleftharpoons Cr(V) + Fe(III)$$

$$Cr(V) + I^- \quad \rightarrow Cr(III) + IO^-$$

$$IO^- + I^- + 2H^+ \rightarrow I_2 + H_2O$$

Whatever the details of the mechanism may be, an intermediate chromium species that reacts more rapidly with iodide ion than does chromate ion itself is formed by the reaction of ferrous ion with chromate ion.

8-5 CHROMATE-ARSENITE REACTION

The oxidation of As(III) by Cr(VI) proceeds with the stoichiometry

$$2Cr(VI) + 3As(III) \rightarrow 2Cr(III) + 3As(V)$$

over a wide range of pH. The probable molecular forms of As(III) in acid and base solutions are $As(OH)_3$ and $AsO(OH)_2^-$, respectively; although there is no question that the forms are neutral monomeric molecule and monomeric anion, respectively, the actual structures are not established. The ionization constant of arsenious acid is 6×10^{-10}. Chromate exists in two molecular forms in acid solution, $Cr_2O_7^=$ and $HCrO_4^-$, and the equilibrium between them

$$2HCrO_4^- \rightleftharpoons Cr_2O_7^= + H_2O$$

has a constant of about fifty. Near neutrality the reaction

$$HCrO_4^- \rightleftharpoons CrO_4^= + H^+$$

occurs, with an ionization constant of 3×10^{-7}.

In view of these equilibria, it is surprising that the reaction can be followed both at the low pH of 1.7 and at the high pH of 10. DeLury[21] studied the kinetics in acid solution; his data (as recomputed by Westheimer[20]) indicate the rate law

$$v = k[As(OH)_3][HCrO_4^-][H^+]^2 \tag{8-11}$$

There is a suggestion of a slight trend toward an increased rate constant at lower acidities.

The Cr(VI)-As(III) reaction in acid solution induces the oxidation of manganous ion to manganese dioxide. Since under the conditions of the experiments the oxidation of Mn(II) by Cr(VI) is thermodynamically impossible, it must be concluded that some intermediate oxidation state of chromium is involved. The induction factor approaches a limiting value of 0.5 as the Mn(II) concentration becomes large in comparison to the As(III) concentration. The fact that two electrons are contributed by arsenic for each one contributed by manganese suggests that the intermediate contains tetravalent chromium.

When As(III) is oxidized by Cr(VI) in the presence of iodide ion, two equivalents of iodine are oxidized for each one of arsenic. The induction factor in this case is therefore 2, suggesting the presence of a pentavalent chromium intermediate.

Westheimer[20] proposes the reaction scheme

$$Cr(VI) + As(III) \rightarrow Cr(IV) + As(V)$$

$$Cr(IV) + Cr(VI) \rightarrow 2Cr(V)$$

$$Cr(V) + As(III) \rightarrow Cr(III) + As(V)$$

when only the two reactants are present. Both of the unusual oxidation states are formed, and in the presence of manganous or iodide ions, they are presumed to react in accord with the steps

$$Cr(IV) + Mn(II) \rightarrow Mn(III) + Cr(III)$$

$$Cr(V) + I^- \rightarrow Cr(III) + IO^-$$

The species Mn(III) and IO^- then go on to form MnO_2 and I_2, respectively.

The chromate-arsenite reaction has also been studied in alkaline solution.[19] At a pH between 9 and 10, the rate is first-order in concentration of each component and zero-order in hydrogen ion concentration. Thus the rate law is

$$v = k[CrO_4^-][AsO(OH)_2^-] \tag{8-12}$$

implying the constitution $(CrAsO_6 \cdot xH_2O)^{3-}$ for the activated complex. This is to be contrasted with the activated complex in acid solution, which has four more protons. It seems probable that other activated complexes with intermediate numbers of protons exist in the pH ranges between those studied.

The reaction in alkaline solution induces the reduction of oxygen with a limiting induction factor of $\frac{4}{3}$. Thus, at the limiting high oxygen pressure, one oxygen molecule is reduced for each chromate ion reduced. Kolthoff and Fineman[19] propose the mechanism

$$Cr(VI) + As(III) \rightarrow Cr(IV) + As(V)$$
$$Cr(IV) + As(III) \rightarrow Cr(II) + As(V)$$
$$Cr(II) + Cr(VI) \rightarrow Cr(III) + Cr(V)$$
$$Cr(V) + As(III) \rightarrow Cr(III) + As(V)$$

for the reactants in the absence of oxygen. The species $Cr(II)$ is assumed in order to provide a particle that is known to react rapidly with oxygen gas. For the additional steps in the induced reactions, they propose

$$Cr(II) + O_2 \rightarrow Cr^2O_2$$
$$2Cr^2O_2 \rightarrow (Cr^2O_2)_2$$
$$(Cr^2O_2)_2 + 2As(III) + 4H^+ \rightarrow 2As(V) + 2Cr(III) + 2OH^- + H_2O_2$$
$$H_2O_2 + As(III) \rightarrow As(V) + 2OH^-$$

with the actual nature of the chromium-oxygen complex Cr^2O_2 unspecified.

8-6   UREA-FORMATION REACTION

The reaction of ammonium ion and cyanate ion to give urea [22]

$$NH_4^+ + NCO^- \rightleftharpoons (H_2N)_2CO$$

proceeds at a measurable rate; it has an equilibrium constant of about 10^4, so the reverse reaction can usually be neglected. The rate law is

$$v = k[NH_4^+][NCO^-] \tag{8-13}$$

and investigators found that added ammonia had no effect on the rate.

On the grounds that this reaction apparently involved the collision of two oppositely charged ions, various electrostatic theories of chemical kinetics were tested. The rate decreased as the ionic strength increased, in the exact manner predicted by the Debye-Hückel theory as applied to ionic rates by Brönsted, Bjerrum, and Christiansen. Also followed was the equation relating rates to dielectric constants, and from this equation came values that were reasonable for the critical distance in the activated complex (about 2×10^{-8} cm).

The excellent correspondence of theory and experiment led many investigators to favor an ionic-collision mechanism. It now seems certain that the actual mechanism is quite different. The problem that arises in the ionic-collision mechanism is the absence of a reasonable way to form a carbon-nitrogen bond when the nitrogen in ammonium ion has four hydrogens about it. This problem finds solution in the equilibrium

$$NH_4^+ + NCO^- \rightleftharpoons NH_3 + HNCO$$

which takes place to a slight extent in aqueous solution.

The preferred mechanism involves nucleophilic attack by molecular ammonia on the carbon of cyanic acid,

$$\underset{H'}{\overset{H}{\underset{\textstyle |}{N}}}-N\text{---}\overset{\overset{\textstyle O}{\|}}{\underset{\underset{\textstyle N}{\|}}{C}} \rightarrow H_3N-C\overset{\textstyle O}{\underset{\textstyle NH}{\diagdown}}$$

followed by rapid redistribution of protons to give the product urea. The evidence for this mechanism may be summarized in these points. First, the objection to the ionic mechanism is overcome. Second, proton redistributions both before and after the rate-determining step are of the types that should occur rapidly. Finally, the rates with different amines when calculated on a molecular basis correlate very well with the rates of reaction of the same amines with carbon dioxide. The latter reactions proceed by a mechanism involving nucleophilic attack by molecular amines on the carbon atom of carbon dioxide.[4]

In retrospect, of course, the electrostatic theories apply to the molecular mechanism just as well as to the ionic mechanism. In both mechanisms, the initial reactants are charged and the activated complex is neutral; the only differences are the path between and the structural detail of the activated complex. It is worth emphasizing here that the electrostatic theories (variation of rate with ionic strength and dielectric constant) deal only with the change in charge from ground state to transition state. Since this change is known independently on the basis of the rate law, the electrostatic theories are of little value in establishing a reaction mechanism.

The other point to be made about this mechanism is that proton redistribution involving oxygen-hydrogen and nitrogen-hydrogen bonds may be important both before and after the rate step. Several equilibria may precede the rate step in order that a transition state of small free-energy of activation may be available to the reactants.

References

1. J. O. Edwards, *Chem. Rev.*, **50**, 455–482 (1952). A compilation of rate laws for oxyanion reactions with donors.
2. J. O. Edwards, *J. Chem. Educ.*, **31**, 270 (1954). A discussion of some factors influencing rates of substitution in oxyanions.
3. G. N. Burkhardt, W. G. K. Ford, and E. Singleton, *J. Chem. Soc.*, **1936**, 17; G. N. Burkhardt, A. G. Evans, and E. Warhurst, *J. Chem. Soc.*, **1936**, 25; G. N. Burkhardt, C. Horrex, and E. I. Jenkins, *J. Chem. Soc.*, **1936**, 1649, 1654.
4. Cf. D. M. Kern, *J. Chem. Educ.*, **37**, 14 (1960). A review on the hydration of carbon dioxide.
5. T. C. Hoering, F. T. Ishimori, and H. O. McDonald, *J. Am. Chem. Soc.*, **80**, 3876 (1958).
6. T. C. Hoering and J. W. Kennedy, *J. Am. Chem. Soc.*, **79**, 56 (1957).
7. C. A. Bunton, E. H. Halevi, and D. R. Llewellyn, *J. Chem. Soc.*, **1952**, 4913, 4917.
8. M. Anbar and S. Guttmann, *J. Am. Chem. Soc.*, **83**, 4741 (1961).
9. T. C. Hoering, R. C. Butler, and H. O. McDonald, *J. Am. Chem. Soc.*, **78**, 4829 (1956).
10. G. A. Mills and H. C. Urey, *J. Am. Chem. Soc.*, **62**, 1019 (1940).
11. M. Anbar and H. Taube, *J. Am. Chem. Soc.*, **76**, 6243 (1954); C. A. Bunton, D. R. Llewellyn, and G. Stedman, *J. Chem. Soc.*, **1959**, 568; C. A. Bunton and H. Masui, *J. Chem. Soc.*, **1960**, 304.
12. M. Anbar and H. Taube, *J. Am. Chem. Soc.*, **80**, 1073 (1958).
13. R. K. Murmann, *J. Inorg. Nucl. Chem.*, **18**, 226 (1961).
14. M. Anbar and S. Guttmann, *J. Am. Chem. Soc.*, **83**, 781 (1961).
15. H. Taube, *Record Chem. Progr.*, **17**, 25 (1956).
16. B. Keisch, J. W. Kennedy, and A. C. Wahl, *J. Am. Chem. Soc.*, **80**, 4778 (1958).
17. C. A. Bunton, D. R. Llewellyn, C. A. Vernon, and V. A. Welch, *J. Chem. Soc.*, **1961**, 1636.
18. "Phosphoric Esters and Related Compounds," *Chem. Soc. (London), Spec. Publ.*, **1957** (No. 8). An interesting survey of organic phosphate mechanisms. See also W. P. Jencks, "Enzyme Models and Enzyme Structure," *Brookhaven Symp. Biol.*, **1962** (No. 15), p. 134.
19. I. M. Kolthoff and M. A. Fineman, *J. Phys. Chem.*, **60**, 1383 (1956).
20. F. H. Westheimer, *Chem. Rev.*, **45**, 419–451 (1949).
21. R. E. DeLury, *J. Phys. Chem.*, **7**, 239 (1903); R. E. DeLury, *J. Phys. Chem.*, **11**, 54 (1907).
22. A review of the kinetics and mechanism data for this reaction is to be found in A. A. Frost and R. G. Pearson, *Kinetics and Mechanism*, 2d ed., Wiley, New York, 1961, p. 307.

9

Free Radical Reactions

A free radical may be defined as a particle with one or more unpaired electrons in either an s or a p orbital. Some radicals are sufficiently stable to be isolated and identified. For example, the species NO_2, NO, ClO_2, O_2, and NF_2 are stable. Pauling[1] ascribes their stability to the formation of a three-electron bond.

Other radicals such as $OH\cdot$, $Cl\cdot$, $Br\cdot$, $H\cdot$, $SO_4^-\cdot$, and $NO_3\cdot$ are too reactive to be isolated. Usually they are formed by homolytic dissociation (thermal or radiation-induced) of weakly bonded molecules. Peroxides, which are known to have a low oxygen-oxygen bond energy, easily dissociate into two oxygen radicals. Symons[2] has reviewed the evidence relevant to the initial stages in the photolysis of hydrogen peroxide and concludes that the steps

$$H_2O_2 + h\nu \rightarrow H_2O_2^*$$

$$H_2O_2^* \rightarrow 2OH\cdot$$

best explain the observations. In the reaction of hydrogen with bromine, the initial radicals result from thermal dissociation of a bromine molecule into bromine atoms.

The mechanisms discussed in this chapter will deal with

inorganic free radicals. The steady-state approximation, some
nonchain reactions, and several chain reactions will be covered.

9-1 STEADY-STATE APPROXIMATION

Free radical reactions usually proceed by mechanisms involving many steps, no one of which is obviously rate-determining. In
chain reactions (see Section 9-4), the initiation step is the step with
the smallest rate constant, but it does not determine the rate law
since the bulk of the product is formed in the propagation steps.
Initiation, propagation, and termination steps all are important
in determining the form of the rate law.

To transform a postulated mechanism into a rate law comparable to that obtained from kinetic experiments, it is useful to
employ an approximation concerning the concentration of
unstable intermediates. This *steady-state approximation*

$$\frac{d[\mathrm{X}]}{dt} \equiv 0 \tag{9-1}$$

where X is an intermediate, says that the concentration of the
intermediate does not change with time. The validity of this
assumption rests upon two conditions: the concentration of the
intermediate must be very small compared to the concentrations
of the reactants and products, and the initial build-up must be
completed. If the intermediate and the reactant concentrations
are comparable, then certainly the assumption is inapplicable.
If the time for initial formation of the intermediate is slow, an
induction period is observed and the steady-state approximation
is invalid until this period is over.

The steady-state approximation is useful for solving mechanism
problems both when the intermediates are radicals and when they
are reactive particles with paired electrons. Its use will be
exemplified by the cases presented in this chapter.

9-2 DECOMPOSITION OF PERSULFATE

Solutions containing peroxydisulfate ion are observed to form
oxygen gas and to become acidic. The reaction

$$\mathrm{S_2O_8^=} + 2\mathrm{H_2O} \rightarrow \mathrm{O_2} + 2\mathrm{SO_4^=} + 4\mathrm{H^+}$$

proceeds at a measurable rate above 50°C. Early workers found it to be first-order in peroxydisulfate ion and to be catalyzed by hydrogen ion; however, their results were somewhat non-reproducible.

The decomposition was carefully reinvestigated by Kolthoff and Miller[3] who observed the rate law

$$v = k_a[S_2O_8^=] + k_b[S_2O_8^=][H^+] \tag{9-2}$$

where the k_b term becomes important only in acidic (pH 2 or less) solutions. By means of oxygen-18 isotope, the source of the oxygen atoms in the product was found to be water for the k_a term and, in part, persulfate for the acid-catalyzed term; therefore, the two terms must relate to different types of mechanism. Since the acid-catalyzed mechanism is poorly understood and probably is non-radical in nature, it will not be discussed here.

The k_a process has been postulated to proceed by a nonchain radical mechanism[3-5] with steps

$$S_2O_8^= \xrightarrow{k_1} 2SO_4^-\cdot$$

$$SO_4^-\cdot + H_2O \xrightarrow{k_2} HSO_4^- + OH\cdot$$

$$2OH\cdot \xrightarrow{k_3} H_2O_2$$

$$H_2O_2 + S_2O_8^= \xrightarrow{k_4} O_2 + 2HSO_4^-$$

$$2H_2O_2 \xrightarrow{k_5} O_2 + 2H_2O$$

$$HSO_4^- \rightleftharpoons H^+ + SO_4^=$$

Applying the steady-state approximation to the intermediates $SO_4^-\cdot$, $OH\cdot$, and H_2O_2, the rate law

$$\frac{-d[S_2O_8^=]}{dt} = 2k_1[S_2O_8^=] \tag{9-3}$$

is obtained when the k_4 step predominates over the k_5 step. If the hydrogen peroxide in step k_5 decomposes spontaneously, as in basic solution, the factor 2 is eliminated from the rate law.

It seems certain that the radical $SO_4^-\cdot$ is formed in the first step of this mechanism. Vinyl monomers (such as styrene in emulsion) are polymerized by small amounts of persulfate with mechanisms consistent with k_1 as an initiation step; furthermore, the resultant polymers are found to have sulfate ion end-groups.

The observed activation energy of 33.5 kcal mole^{-1} for persulfate decomposition is about that expected for homolytic scission of the oxygen-oxygen bond in the rate step.

Evidence supporting this mechanism is found in the work of Tsao and Wilmarth[6] on the photolytic decomposition of peroxydisulfate. When the ion is irradiated in aqueous solution at room temperature with 2537-A light, the reaction

$$S_2O_8^= + H_2O \rightarrow 2HSO_4^- + \tfrac{1}{2}O_2$$

occurs with a quantum yield of about 0.6. The presence of hydrogen ions and sulfate ions together strongly inhibits the photolytic reaction, and the quantum yield decreases to about 20 per cent of its uninhibited value. The path responsible for the 80 per cent of the photolytic reaction that can be inhibited is accompanied by exchange of sulfur isotopes between added labeled sulfate ion and peroxydisulfate. As in the thermal decomposition, the atoms in the product oxygen molecules are derived from solvent water; therefore, both the thermal and the photolytic decompositions are oxidations of water by peroxydisulfate ion.

A reaction scheme consistent with these facts is

$$h\nu + S_2O_8^= \rightarrow 2SO_4^-\cdot$$

$$SO_4^-\cdot + H_2O \rightleftharpoons OH\cdot + H^+ + SO_4^=$$

$$2SO_4^-\cdot \rightarrow S_2O_8^=$$

$$2OH\cdot \rightarrow H_2O_2$$

with hydrogen peroxide going to oxygen as before.

The second step in this scheme appears to be rapidly reversible despite the short lifetimes of the two radicals. Thus, this step provides a means for the rapid exchange of sulfur isotope between the sulfate ion and the sulfate ion radical $SO_4^-\cdot$. The third step then returns the tagged atom, which originated in sulfate ion, to the peroxydisulfate. This reaction scheme also shows how the quantum yield for decomposition would be lowered by the combination of hydrogen and sulfate ions. These differences between the thermal and photolytic decompositions result from the higher steady-state concentration of the sulfate ion radical in the photolytic process.

Little is known about the follow-up steps k_3 through k_5 in the two decompositions. The presence of hydrogen peroxide has not been detected, but a difference by a factor of two in the observed rate constants has been found,[5] depending on whether k_4 or k_5 is the final step.

9-3 DINITROGEN PENTOXIDE DECOMPOSITION

During the early days of chemical kinetics, the gas-phase reaction

$$2N_2O_5 \rightarrow 4NO_2 + O_2$$

$$2NO_2 \rightleftharpoons N_2O_4$$

was found to be a first-order decomposition. In the hope that it would have a simple, clean-cut, unimolecular rate step and thereby serve as an adequate test of the collision theory of kinetics, it was thoroughly investigated. The hope, unfortunately, was not borne out, for the reaction proved to be complex. The presently accepted mechanism is that postulated by Ogg.[7]

$$N_2O_5 \xrightarrow{k_1} NO_3 + NO_2$$

$$NO_2 + NO_3 \xrightarrow{k_2} N_2O_5$$

$$NO_2 + NO_3 \xrightarrow{k_3} NO + O_2 + NO_2$$

$$NO + NO_3 \xrightarrow{k_4} 2NO_2$$

$$2NO_2 \rightleftharpoons N_2O_4$$

The differential equation for concentration of the unstable radical NO_3 is

$$\frac{d[NO_3]}{dt} = k_1[N_2O_5] - k_2[NO_2][NO_3] - k_3[NO_2][NO_3]$$
$$- k_4[NO][NO_3] \tag{9-4}$$

and that for the intermediate NO is

$$\frac{d[NO]}{dt} = k_3[NO_2][NO_3] - k_4[NO][NO_3] \tag{9-5}$$

Setting the latter equal to zero and rearranging,

$$[NO] = \frac{k_3}{k_4}[NO_2] \tag{9-6}$$

On the basis of the steady-state approximation [Eq. (9-1)] and Eq. (9-6) for [NO], Eq. (9-4) may be solved to give

$$[NO_3] = \frac{k_1[N_2O_5]}{(k_2 + 2k_3)[NO_2]} \tag{9-7}$$

From the equation for product oxygen formation, we have

$$\frac{d[O_2]}{dt} = k_3[NO_2][NO_3] \tag{9-8}$$

Therefore

$$\frac{d[O_2]}{dt} = \frac{k_1 k_3[N_2O_5]}{k_2 + 2k_3} \tag{9-9}$$

Since k_2 probably has no activation energy whereas the k_3 step does, k_3 must be much smaller than k_2 and the rate law from the postulated mechanism reduces to

$$\frac{d[O_2]}{dt} = \frac{k_1 k_3}{k_2}[N_2O_5] \tag{9-10}$$

so that

$$k_{obs} = \frac{k_1 k_3}{k_2} \tag{9-11}$$

In spite of the simplicity of the observed rate law, the first-order constant is actually a product of several constants. The ratio k_1/k_2 is an equilibrium constant and k_3 is the rate constant for a bimolecular reaction; therefore the observed rate constant is not expected to and does not follow the predictions of theory for unimolecular reactions.

There are several aspects of this mechanism that can be subjected to experimental verification.[8] The first two steps of the mechanism, if rapid as Ogg[7] suggests, constitute a path for nitrogen-isotope exchange between NO_2 and N_2O_5; also, if the first step is rate-determining as expected, the rate should be first-order in pentoxide and zero-order in dioxide. Both phenomena have been observed in the gas phase and in carbon tetrachloride solutions. Furthermore, the gas-phase rate decreases with pressure in agreement with the theory of unimolecular reactions.

The reaction of nitric oxide with the pentoxide

$$NO + N_2O_5 \rightarrow 3NO_2$$

also provides a test of the decomposition mechanism. According to the mechanism, this reaction should proceed as follows

$$N_2O_5 \rightleftharpoons NO_2 + NO_3$$

$$NO_3 + NO \rightarrow 2NO_2$$

with the first step being the slow step. It has been found to be first-order in pentoxide and zero-order in nitric oxide, with some inhibition by nitrogen dioxide. At the same total pressure, the rate of this reaction is identical to the rate of the isotope exchange mentioned above.

The activation energy for the decomposition of N_2O_5 has been found to be 24.6 kcal mole^{-1}; this compares favorably with the value 24.3 kcal mole^{-1}, the calculated endothermicity of the process

$$N_2O_5 \rightarrow NO_2 + O_2 + NO$$

Evidently there is little or no energy barrier other than the endothermicity of the first and third steps.

One of the interesting points about the decomposition of dinitrogen pentoxide is its insensitivity to foreign substances. The reaction is presumed to be homogeneous, since neither the nature nor the size of the wall surfaces has any effect on the rate constant. This agrees with the previously mentioned fact that the rate constant is only slightly affected by the medium. Formally speaking, the ratio between reactant and activated-complex activity coefficients must be insensitive to the medium.

9-4 CHAIN REACTIONS

Certain free radical reactions proceed by mechanisms in which the bulk of the product is formed through cyclic steps involving neither gain nor loss of radicals. These steps are called *propagation steps* or chain-carrying steps. The radicals acting as chain carriers are formed in the *initiation step* and are destroyed in the *termination step*. These three types of steps are characteristic of all chain reactions, and the kinetics observed depend on the nature of these steps.

The oxidation of 2-propanol by peroxydisulfate[9]

$$S_2O_8^= + HC(CH_3)_2OH \rightarrow 2HSO_4^- + C(CH_3)_2O$$

is an excellent example of a chain reaction. In the absence of oxygen and other inhibitors, the kinetics are

$$v = k_{obs}[S_2O_8^=][HC(CH_3)_2OH]^{\frac{1}{2}} \tag{9-12}$$

with a rate constant of 8×10^{-3} liter$^{\frac{1}{2}}$ mole$^{-\frac{1}{2}}$ sec^{-1} at 50°C and with an activation energy of 21 kcal mole^{-1}. The initiation step presumably is

$$S_2O_8^= \xrightarrow{k_1} 2SO_4^-\cdot$$

Since the rate of peroxydisulfate loss in the isopropyl alcohol solution is reduced by the addition of allyl acetate essentially to the known decomposition rate of peroxydisulfate in pure water, the alcohol oxidation and the aqueous decomposition must be initiated by the same step.

The postulated propagation steps are

$$SO_4^-\cdot + HC(CH_3)_2OH \xrightarrow{k_2} HSO_4^- + \cdot C(CH_3)_2OH$$

$$\cdot C(CH_3)_2OH + S_2O_8^= \xrightarrow{k_3} C(CH_3)_2O + HSO_4^- + SO_4^-\cdot$$

Evidence for the existence of the radical $\cdot C(CH_3)_2OH$ is provided by the observation that oxygen gas is a powerful inhibitor. The oxygen molecule, which is in the triplet electronic state, reacts very rapidly with organic radicals; this process is essentially a radical combination. Addition of the two propagation steps gives

$$S_2O_8^= + HC(CH_3)_2OH \rightarrow 2HSO_4^- + C(CH_3)_2O$$

which is the stoichiometry of the over-all reaction. Since the propagation steps constitute a self-contained cycle, a large number of reactant molecules can be turned into products every time one initiation process transpires. Under the conditions of the study cited here,[9] it was observed that 1800 molecules of peroxydisulfate were used up in the propagation steps for each one lost in an initiation step. Thus the *chain length* for the reaction is 1800.

Radicals recombine at nearly every collision since the termination steps have essentially zero activation energy. For the persulfate-isopropanol reaction, the postulated termination step is

$$SO_4^-\cdot + \cdot C(CH_3)_2OH \xrightarrow{k_4} HSO_4^- + C(CH_3)_2O$$

with one of each type of radical reacting to give nonradical products.

As a consequence of the steady-state condition in a chain reaction, the rate of initiation must equal the rate of termination; therefore

$$k_1[S_2O_8^=] = k_4[SO_4^-\cdot][\cdot C(CH_3)_2OH] \qquad (9\text{-}13)$$

We also know that

$$k_2[SO_4^-\cdot][HC(CH_3)_2OH] = k_3[\cdot C(CH_3)_2OH][S_2O_8^=] \qquad (9\text{-}14)$$

These two equations, which follow from the differential equations for the concentrations of the two radicals and from the steady-state approximation [Eq. (9-1)], can be solved to obtain expressions [Eqs. (9-15) and (9-16)]

$$[SO_4^-\cdot] = \left(\frac{k_1 k_3}{k_2 k_4}\right)^{\frac{1}{2}} \frac{[S_2O_8^=]}{[HC(CH_3)_2OH]^{\frac{1}{2}}} \qquad (9\text{-}15)$$

$$[\cdot C(CH_3)_2OH] = \left(\frac{k_1 k_2}{k_3 k_4}\right)^{\frac{1}{2}} [HC(CH_3)_2OH]^{\frac{1}{2}} \qquad (9\text{-}16)$$

for the concentrations of the radicals in terms of stable species. When Eq. (9-16) is applied to Eq. (9-17),

$$\frac{-d[S_2O_8^=]}{dt} = k_1[S_2O_8^=] + k_3[\cdot C(CH_3)_2OH][S_2O_8^=] \qquad (9\text{-}17)$$

there results Eq. (9-18).

$$\frac{-d[S_2O_8^=]}{dt} = k_1[S_2O_8^=] + \left(\frac{k_1 k_2 k_3}{k_4}\right)^{\frac{1}{2}} [S_2O_8^=][HC(CH_3)_2OH]^{\frac{1}{2}}$$
$$(9\text{-}18)$$

The first term can be neglected since it corresponds to the initiation step, which is known from the chain length to contribute less than 0.1 per cent of the total reaction; therefore, the rate law obtained from the postulated chain mechanism, Eq. (9-19),

$$\frac{-d[S_2O_8^=]}{dt} = \left(\frac{k_1 k_2 k_3}{k_4}\right)^{\frac{1}{2}} [S_2O_8^=][HC(CH_3)_2OH]^{\frac{1}{2}} \qquad (9\text{-}19)$$

agrees with the experimental rate law.

It is seen that

$$k_{obs} = \left(\frac{k_1 k_2 k_3}{k_4}\right)^{\frac{1}{2}} \qquad (9\text{-}20)$$

from which it follows that the activation energy E_a for k_{obs} is

$$E_a = \tfrac{1}{2}(E_1 + E_2 + E_3 - E_4) \qquad (9\text{-}21)$$

The value of E_a is 21 kcal mole^{-1}; that for E_1, which can be obtained from the peroxydisulfate decomposition, is 33.5; and it is a reasonable postulate that $E_4 \simeq 0$. Thus $E_2 + E_3 \simeq 8$ kcal mole^{-1}. The termination step is a cross-termination in this case; that is, one of each of the two radicals is involved. Assuming that $E_2 \simeq E_3$, each has a value of roughly 4 kcal mole^{-1}. This is reasonable because it is within the range of values for processes involving hydrogen transfer via radical displacement.[10]

The form of the rate law for a chain reaction depends on the radicals involved in the termination step. This dependence may be easily confirmed by setting up a chain mechanism and holding everything constant but the termination step; the rate law obtained will vary in a manner that becomes predictable with experience.

Since termination steps for homogeneous reactions are radical recombinations, they all have about the same rate constant. The termination step that is observed in a particular case will involve the most stable radical present, for the most stable radical builds up to a higher steady-state concentration than the others and is, therefore, more likely to be present in a termination step. For example, in autoxidation reactions, the termination step usually involves two peroxy radicals

$2ROO \cdot \rightarrow$ products

Since peroxy radicals are nonreactive and have relatively long lifetimes, there is a good chance for two such radicals to meet. Conversely, extremely reactive radicals such as hydrogen atoms will react with stable molecules before they have much chance to diffuse together and combine.

9-5 HYDROGEN-BROMINE REACTION

In 1907 Bodenstein and Lind[11] studied the reaction

$H_2 + Br_2 \rightarrow 2HBr$

in the temperature range 228 to 303°C. The rate law was found to be

$$\frac{d[HBr]}{dt} = \frac{k[H_2][Br_2]^{\frac{1}{2}}}{1 + (k'[HBr]/[Br_2])} \tag{9-22}$$

with an activation energy for k of 40.2 kcal mole^{-1}. The constant k' is temperature-independent and has a value of 0.1. The rate law was explained simultaneously by Christiansen,[12] Herzfeld,[13] and Polanyi[14] in terms of the mechanism

$$Br_2 \xrightarrow{k_1} 2Br\cdot$$

$$Br\cdot + H_2 \xrightarrow{k_2} HBr + H\cdot$$

$$H\cdot + Br_2 \xrightarrow{k_3} HBr + Br\cdot$$

$$H\cdot + HBr \xrightarrow{k_4} H_2 + Br\cdot$$

$$2Br\cdot \xrightarrow{k_5} Br_2$$

The k_1 step is an initiation step; the k_2, k_3, and k_4 steps are termed the propagation steps although the latter one regenerates the reactant hydrogen molecule. The termination step is k_5. It is possible to write five differential equations, one for each of the chemical species present. Those for radical concentrations, Eqs. (9-23) and (9-24),

$$\frac{d[Br\cdot]}{dt} = 2k_1[Br_2] - k_2[Br\cdot][H_2] + k_3[H\cdot][Br_2] +$$
$$k_4[H\cdot][HBr] - 2k_5[Br\cdot]^2 \quad (9\text{-}23)$$

$$\frac{d[H\cdot]}{dt} = k_2[Br\cdot][H_2] - k_3[H\cdot][Br_2] - k_4[H\cdot][HBr] \quad (9\text{-}24)$$

are each equal to zero, since the steady-state approximation [Eq. (9-1)] is presumed to hold for the bromine and hydrogen atoms. Adding Eqs. (9-23) and (9-24), we find that

$$2k_1[Br_2] - 2k_5[Br\cdot]^2 = 0 \quad (9\text{-}25)$$

from which is obtained

$$[Br\cdot] = (k_1/k_5)^{\frac{1}{2}}[Br_2]^{\frac{1}{2}} \quad (9\text{-}26)$$

Substitution into the equation for hydrogen atom concentration [Eq. (9-24)] yields the equality

$$[H\cdot] = \frac{k_2(k_1/k_5)^{\frac{1}{2}}[H_2][Br_2]^{\frac{1}{2}}}{k_3[Br_2] + k_4[HBr]} \quad (9\text{-}27)$$

These expressions for atom concentrations [Eqs. (9-26) and (9-27)] may then be substituted into the differential equation for product formation to give

$$\frac{d[HBr]}{dt} = \frac{2k_2(k_1/k_5)^{\frac{1}{2}}[H_2][Br_2]^{\frac{1}{2}}}{1 + (k_4[HBr]/k_3[Br_2])} \quad (9\text{-}28)$$

This rate law, derived from the postulated mechanism, agrees with the observed rate law when $k = 2k_2 (k_1/k_5)^{\frac{1}{2}}$ and $k' = k_4/k_3$.

It is of interest to compare the experimental activation energies with the known thermodynamic quantities:

$$Br_2 \rightleftharpoons 2Br\cdot \qquad\qquad E^0(0°C) = 45.4 \text{ kcal mole}^{-1}$$

$$H_2 + Br\cdot \rightleftharpoons HBr + H\cdot \qquad E^0(0°C) = 17.0 \text{ kcal mole}^{-1}$$

$$H\cdot + Br_2 \rightleftharpoons HBr + Br\cdot \qquad E^0(0°C) = -41.1 \text{ kcal mole}^{-1}$$

The over-all activation energy is related to the activation energies of the individual steps in accord with Eq. (9-29)

$$E_a = E_2 + \tfrac{1}{2}E_1 - \tfrac{1}{2}E_5 \tag{9-29}$$

where E_2 is the activation energy for the k_2 step, etc. E_1 minus E_5 must equal 45.4 kcal mole^{-1}, since the difference between activation energies of initiation and termination is the heat of dissociation of the bromine molecule. It follows, therefore, that

$$E_2 = E_a - \tfrac{1}{2}(45.4) = 17.5 \text{ kcal mole}^{-1}$$

which is comparable with the endothermicity of 17.0 for this step.

The fact that the constant k' is temperature-independent is not surprising since it is a ratio between two constants for exothermic propagation steps. Since both steps should have small and similar temperature dependences, the ratio is not expected to change with temperature.

The chain nature of this reaction was confirmed by photo-initiation studies.[15-17] Over a range of pressures, the experimental rate law is

$$\frac{d[HBr]}{dt} = \frac{k\nu[H_2]\,I_{\text{abs}}^{\frac{1}{2}}}{P^{\frac{1}{2}}\{1 + (k'[HBr]/[Br_2])\}} \tag{9-30}$$

where I_{abs} is the intensity of the light absorbed and P is the total pressure. Since the bromine molecule is the only species that absorbs light, the initiation step now becomes

$$Br_2 + h\nu \xrightarrow{k'_1} 2Br\cdot$$

and the termination step

$$2Br\cdot + M \xrightarrow{k'_5} Br_2 + M$$

Thus the steady-state concentration of the bromine atom is

$$[Br\cdot] = \left(\frac{k'_1 I_{\text{abs}}}{k'_5[M]}\right)^{\frac{1}{2}} \tag{9-31}$$

(It was not necessary to use the third body M in the earlier mechanism, since both initiation and termination were equally influenced and [M] cancelled out in the ratio.) On the basis of the steady-state theory and the same propagation steps as in the thermal reaction, the derived rate law is

$$\frac{d[\text{HBr}]}{dt} = \frac{2k_2(k_1'/k_5')^{\frac{1}{2}}[\text{H}_2]\, I_{\text{abs}}^{\frac{1}{2}}}{[\text{M}]^{\frac{1}{2}}\{1 + (k_4[\text{HBr}]/k_3[\text{Br}_2])\}} \tag{9-32}$$

which agrees with the experimental rate law [Eq. (9-30)] when it is assumed that [M] is proportional to P.

The activation energy for the photochemical reaction was found to be 17.6 kcal mole^{-1}. Since the k_1' and k_5' steps should be nearly temperature-independent, this energy should and does compare well with that calculated for the k_2 step of the thermal reaction.

The reaction of hydrogen and bromine was studied carefully by Bodenstein and his co-workers, and it is still studied today. Recently it was found[18] in shock-wave kinetic studies that the hydrogen molecule formed in the reaction

$$\text{H}\cdot + \text{HBr} \rightarrow \text{H}_2 + \text{Br}\cdot$$

is in the first excited vibrational level. It was concluded[18] that the rate of a chemical reaction may vary, depending on which quantum state of the products is formed. In the shock experiments, the rates of dissociation of HBr between 2000 and 3000°K agreed with values of rate constants extrapolated from the previous work.

9-6 OTHER HYDROGEN-HALOGEN REACTIONS

Although all of the hydrogen-halogen reactions have some mechanistic similarities, there are also differences. These are particularly striking in the hydrogen-iodine reaction, which at lower temperatures proceeds primarily by a nonradical mechanism. The rate law is

$$v = k[\text{H}_2][\text{I}_2] \tag{9-33}$$

and the postulated mechanism involves the simultaneous breaking of two bonds and formation of two bonds in an activated complex of structural type (I). The activation energy in the forward direction is 40.0 kcal mole^{-1} and in the reverse direction 44.0 kcal mole^{-1}; the entropies of activation for both directions are near zero. The early work on this reaction was done by Bodenstein.[19]

(I)

At higher temperatures, a radical process analogous to the bromine case becomes important in the formation of hydrogen iodide. Recent careful work[20,21] has indicated that this radical process, which has an activation energy of about 51 kcal mole^{-1}, produces only about 10 per cent of the total hydrogen iodide formed at 360°C but about 95 per cent of the product formed at 530°C. The activation energy for the radical mechanism is high in spite of the low bond-dissociation energy of the iodine molecule; the reason is that the endothermicity of the step

$$I\cdot + H_2 \rightarrow HI + H\cdot$$

amounts to 32.8 kcal mole^{-1}. Since the activation energy for any primary step is at least as large as the endothermicity, this step will have an activation energy of at least 32.8 kcal mole^{-1}. The ratio between the rate constants for the exothermic steps

$$H\cdot + HI \rightarrow H_2 + I\cdot$$

and

$$H\cdot + I_2 \rightarrow HI + I\cdot$$

has been found to be 0.08 and independent of temperature over a wide range. Evidence indicating that the activation energies for these two steps are actually zero has been obtained; this is consistent with the fact that both steps are exothermic radical displacements.

The radical mechanism becomes predominant in spite of its higher activation energy because it has a higher entropy of activation than the nonradical one. This difference in entropies is caused by the positive entropy for the reaction

$$I_2 \rightleftharpoons 2I$$

which is the initiation step in the radical mechanism.

The hydrogen-chlorine reaction becomes appreciable above 200°C; however, the kinetic results are not completely conclusive. Initiation by light and other types of radiation is important, as is inhibition by oxygen. Furthermore, the rate appears to depend on the nature of the vessel walls, although the surface-to-volume ratio is not invariably important.[22] The evidence is consistent with a free radical chain mechanism involving hydrogen and chlorine atoms as the chain-carrying intermediates, along with a variety of steps involving oxygen molecules.

It is known that absorption of light of any wavelength below 478 mμ by the chlorine molecule will cause dissociation of the molecule into atoms. Thus it seems certain that the photochemical hydrogen-chlorine reaction is initiated by the step

$$Cl_2 + h\nu \rightarrow 2Cl\cdot$$

and that the follow-up steps are analogous to those of the hydrogen-bromine reaction. Proof for the presence of hydrogen atoms in the reaction mixture is obtained through the observation that some ortho-para hydrogen conversion occurs during the reaction of hydrogen with chlorine. This conversion is known to occur via the mechanism

$$H\cdot + H_2(para) \rightarrow H\cdot + H_2(ortho)$$

As in the case of the thermal reaction, the photoinitiated reaction is inhibited by oxygen gas. The observed[23,24] rate law is

$$\frac{d[HCl]}{dt} = \frac{k[H_2][Cl_2]I_{abs}}{m[Cl_2] + [O_2]([H_2] + 0.1[Cl_2])} \tag{9-34}$$

where I_{abs} is the intensity of the light absorbed. The postulated mechanism is

$$Cl_2 + h\nu \overset{k_1}{\rightarrow} 2Cl\cdot$$

$$Cl\cdot + H_2 \overset{k_2}{\rightarrow} HCl + H\cdot$$

$$H\cdot + Cl_2 \overset{k_3}{\rightarrow} HCl + Cl\cdot$$

$$H\cdot + O_2 \overset{k_4}{\rightarrow} HO_2\cdot$$

$$Cl\cdot + O_2 \overset{k_5}{\rightarrow} ClO_2\cdot$$

$$Cl\cdot + Y \overset{k_6}{\rightarrow} ClY$$

where Y is any substance (wall, gas particle, etc.) that can remove chlorine atoms. Using the steady-state approximation [Eq. (9-1)], the rate law expressed by Eq. (9-35)

$$\frac{d[\text{HCl}]}{dt} = \frac{(2k_3/k_4)\,[\text{H}_2]\,[\text{Cl}_2]\,I_{\text{abs}}}{\dfrac{k_3 k_6}{k_2 k_4}[\text{Cl}_2]\,[\text{Y}] + [\text{O}_2]\left([\text{H}_2] + \dfrac{k_3 k_5}{k_2 k_4}[\text{Cl}_2] + k_6/k_2[\text{Y}]\right)}$$

$$(9\text{-}35)$$

can be derived from the mechanism. Apart from the final term in the denominator, the derived rate law and the observed law [Eq. (9-34)] are of the same form.

It is of interest to compare predicted activation energies for the three halogen-hydrogen reactions with the observed mechanisms. The data are given in Table 9-1. The calculated energies for the bimolecular mechanism are from Laidler's book,[25] and the energies for the radical steps are in each case zero or the endothermicity of the reaction. Although the calculated values for activation energies of the bimolecular process may be somewhat in error, it is probable that the similarity in values for these three halogens is real. For the radical process, the complete calculation of the activation energy is shown in the last line of the table. It can be seen that the over-all energy order Cl < Br < I is established primarily by the energy of the process

$$\text{X}\cdot + \text{H}_2 \rightarrow \text{HX} + \text{H}\cdot$$

Table 9-1 *Calculated activation energies for the hydrogen-halogen reactions*[a]

	Halogen X		
Reaction	Cl	Br	I
$\text{H}_2 + \text{X}_2 \rightarrow 2\text{HX}$	50	45	48[b]
$\frac{1}{2}\text{X}_2 \rightleftharpoons \text{X}\cdot$	28[c]	23[c]	18[c]
$\text{X}\cdot + \text{H}_2 \rightarrow \text{HX} + \text{H}\cdot$	8	17	33
$\text{H}\cdot + \text{X}_2 \rightarrow \text{HX} + \text{X}\cdot$	0	0	0
Sum of last three values	36	40[d]	51[d]

[a] All values in kcal mole^{-1}.
[b] Observed value is 40.
[c] Half the heat of dissociation of the halogen molecule.
[d] These values have also been observed.

which is slightly endothermic (8 kcal mole^{-1}) in the case of chlorine but strongly endothermic (33 kcal mole^{-1}) in the case of iodine.

For the hydrogen-iodine reaction, then, the low-activation-energy nonradical path predominates at low temperatures. At higher temperatures, the radical path of higher activation energy and higher frequency factor takes over. In the case of the hydrogen-chlorine reaction, the radical reaction will probably be predominant at all temperatures because of its much lower activation energy. Only a radical path for the hydrogen-bromine reaction has been observed to date; however, the similarity in calculated energies for the competing mechanisms suggests that a careful search for a nonradical process might be successful.

The hydrogen-fluorine reaction probably has an atomic mechanism similar to that of chlorine and bromine except that the chain is more easily interrupted.[26] Since the heat of dissociation of the fluorine molecule is only 37 kcal mole^{-1} and since both of the chain-propagation steps are exothermic, the chain mechanism should have an activation energy of only about 19 kcal mole^{-1}.

9-7 PERSULFATE-PEROXIDE REACTION

In an effort to learn more about oxygen radicals, Tsao and Wilmarth[27] investigated the stoichiometry and kinetics of the interaction of peroxydisulfate ion and hydrogen peroxide in aqueous solution. Under certain conditions, the stoichiometry is

$$S_2O_8^= + H_2O_2 \rightarrow O_2 + 2HSO_4^-$$

The kinetics were found to be quite complicated, for different rate laws assume importance as concentrations are varied; this is one indication that the mechanism involves free radicals. With three different ranges of reactant concentrations, the rate law approaches three different limiting forms: first-order in peroxydisulfate and half-order in hydrogen peroxide; first-order in peroxydisulfate; and half-order in peroxydisulfate. Evidence was also found for a rate law that is half-order in peroxydisulfate and half-order in hydrogen peroxide; however, at no level of concentration did this rate law represent a major portion of the total rate of reaction. The complete empirical rate law was written in the form

$$\frac{-d[S_2O_8^=]}{dt}$$

$$= \frac{[S_2O_8^=]}{\left(\dfrac{9.5 \times 10^6}{[H_2O_2]} + \dfrac{(7.6 \times 10^7)\,[S_2O_8^=]}{[H_2O_2]} + 7.9 \times 10^8 + (2.9 \times 10^{10})\,[S_2O_8^=]\right)^{\frac{1}{2}}}$$

$$(9\text{-}36)$$

This rate law has been rationalized in terms of a free radical chain mechanism where four different termination steps are important, depending on the reactant concentrations. The postulated initiation step is the unimolecular homolytic scission of peroxydisulfate ion

$$S_2O_8^= \xrightarrow{k_1} 2SO_4^-\cdot$$

and the chain-propagation steps are

$$SO_4^-\cdot + H_2O \underset{k_3}{\overset{k_2}{\rightleftharpoons}} HSO_4^- + OH\cdot$$

$$OH\cdot + H_2O_2 \xrightarrow{k_4} H_2O + HO_2\cdot$$

$$HO_2\cdot + S_2O_8^= \xrightarrow{k_5} HSO_4^- + O_2 + SO_4^-\cdot$$

The three radicals present, $OH\cdot$, $SO_4^-\cdot$, and $HO_2\cdot$, can each take part in termination steps. Therefore the six possible termination steps are

$$HO_2\cdot + OH\cdot \xrightarrow{k_a} H_2O + O_2$$

$$SO_4^-\cdot + OH\cdot \xrightarrow{k_b} HSO_5^-$$

$$HO_2\cdot + SO_4^-\cdot \xrightarrow{k_c} O_2 + HSO_4^-$$

$$SO_4^-\cdot + SO_4^-\cdot \xrightarrow{k_d} S_2O_8^=$$

$$HO_2\cdot + HO_2\cdot \xrightarrow{k_e} H_2O_2 + O_2$$

$$OH\cdot + OH\cdot \xrightarrow{k_f} H_2O_2$$

Of these six steps, the last two may be disregarded since the first four satisfy the observed rate law. Using the steady-state approximation

[Eq. (9-1)] and neglecting the rates of chain initiation and termination as compared to the rate of chain propagation, one can derive the following rate law:

$$\frac{-d[S_2O_8^=]}{dt}$$

$$= \frac{[S_2O_8^=]}{\left(\dfrac{k_a}{k_1 k_4 k_5 [H_2O_2]} + \dfrac{k_b[S_2O_8^=]}{k_1 k_2 k_4 [H_2O_2]} + \dfrac{k_c}{k_1 k_2 k_5} + \dfrac{k_d[S_2O_8^=]}{2k_1 k_2^2}\right)^{\frac{1}{2}}}$$

$$(9\text{-}37)$$

This derived law has the same form as the observed law [Eq. (9-36)], and it is possible to compare the complex rate-constant formulations with the experimental quantities, as, for example,

$$\frac{k_c}{k_1 k_2 k_5} = 7.9 \times 10^8$$

Although there is good agreement between the observed and derived rate laws, several questions arise. Why should k_d be important in this reaction but not in the thermal-decomposition reaction where the $SO_4^-\cdot$ concentration is probably greater? Also, since the radical $HO_2\cdot$ is expected to be the most stable radical present, why is the k_e termination step not observed to be important?

When hydrogen peroxide is present in large excess, the stoichiometry is a mixture of the previous one and that for hydrogen peroxide decomposition

$$2H_2O_2 \rightarrow 2H_2O + O_2$$

Along with the propagation constants k_2 through k_5 mentioned above, a new one must be considered:

$$HO_2\cdot + H_2O_2 \xrightarrow{k_6} O_2 + H_2O + OH\cdot$$

The radical $HO_2\cdot$ can react with either $S_2O_8^=$ or H_2O_2. As the concentration of hydrogen peroxide increases, the k_6 step will predominate over k_5 and a greater fraction of hydrogen peroxide will be lost by decomposition.

References

1. L. C. Pauling, *The Nature of the Chemical Bond*, 3d ed., Cornell University Press, Ithaca, N.Y., 1960.
2. M. C. R. Symons, in J. O. Edwards (ed.), *Peroxide Reaction Mechanisms*, Wiley-Interscience, New York, 1962, Chap. 8, p. 137ff.

3. I. M. Kolthoff and I. K. Miller, *J. Am. Chem. Soc.*, **73**, 3055 (1951); earlier studies are referenced in this paper.

4. P. D. Bartlett and J. D. Cotman, Jr., *J. Am. Chem. Soc.*, **71**, 1419 (1949).

5. W. K. Wilmarth and A. Haim, in J. O. Edwards (ed.), *Peroxide Reaction Mechanisms*, Wiley-Interscience, New York, 1962, Chap. 10, p. 175ff. An excellent review of the mechanisms for peroxodisulfate reactions.

6. M. S. Tsao and W. K. Wilmarth, *J. Phys. Chem.*, **63**, 346 (1959).

7. R. A. Ogg, Jr., *J. Chem. Phys.*, **15**, 337 (1947); **18**, 572 (1950).

8. A. A. Frost and R. G. Pearson, *Kinetics and Mechanism*, 2d ed., Wiley, New York, 1961, p. 380ff. A discussion in some detail of the mechanism of the dinitrogen pentoxide decomposition.

9. D. L. Ball, M. M. Crutchfield, and J. O. Edwards, *J. Org. Chem.*, **25**, 1599 (1960).

10. A. A. Frost and R. G. Pearson, *Kinetics and Mechanism*, 2d ed., Wiley, New York, 1961, Chap. 6.

11. M. Bodenstein and S. C. Lind, *Z. Physik. Chem. (Leipzig)*, **57**, 168 (1907).

12. J. A. Christiansen, *Kgl. Danske Videnskab. Selskab., Mat.-Fys. Medd.*, **1**, 14 (1919).

13. K. F. Herzfeld, *Ann. Physik*, **59**, 635 (1919).

14. M. Polanyi, *Z. Elektrochem.*, **26**, 49 (1920).

15. M. Bodenstein and H. Lutkemeyer, *Z. Physik. Chem. (Leipzig)*, **114**, 108 (1925).

16. M. Bodenstein and G. Jung, *Z. Physik. Chem. (Leipzig)*, **121**, 127 (1926).

17. W. Jost and G. Jung, *Z. Physik. Chem.*, **B3**, 83 (1929).

18. R. J. Araujo, Ph.D. thesis, Brown University, 1962.

19. M. Bodenstein, *Z. Physik. Chem. (Leipzig)*, **13**, 56 (1894); **22**, 1 (1897); **29**, 295 (1899).

20. S. W. Benson and R. Srinivasan, *J. Chem. Phys.*, **23**, 200 (1955).

21. J. H. Sullivan, *J. Chem. Phys.*, **30**, 1292, 1577 (1959); **36**, 1925 (1962).

22. R. N. Pease, *J. Am. Chem. Soc.*, **56**, 2388 (1934).

23. N. Thon, *Z. Physik. Chem. (Leipzig)*, **124**, 327 (1926).

24. M. Bodenstein and W. Unger, *Z. Physik. Chem.*, **B11**, 253 (1930).

25. K. J. Laidler, *Chemical Kinetics*, McGraw-Hill, New York, 1950, p. 217.

26. M. Bodenstein, H. Jockusch, and S.-H. Chong, *Z. Anorg. Allgem. Chem.*, **231**, 24 (1937).

27. M.-S. Tsao and W. K. Wilmarth, *Discussions Faraday Soc.*, **1960** (No. 29), 137; M.-S. Tsao and W. K. Wilmarth, paper presented at ACS National Meeting, Atlantic City, September, 1962.

10

Some Unsolved Problems

The many types of inorganic reaction mechanisms make it difficult to classify or in some cases even to describe them. The difficulty stems partly from our limited knowledge of certain areas of inorganic chemistry. This chapter will deal in general and specific terms with some of these yet unsolved problems.

10-1 GENERAL PROBLEMS

Surprisingly, the molecular structures of many inorganic compounds and ions are still not established. The iodate ion, which as IO_3^- should have a pyramidal configuration like ClO_3^-, has been suggested to exist in water in the form $H_2IO_4^-$. The chromous ion in water may be octahedrally coordinated, or it may have tetragonal symmetry with four water molecules in a plane close to the metal and with two other water molecules *trans* to each other but further away from the metal ion. The extent of solvation of square-planar complexes like $[Cu(NH_3)_4]^{++}$ and $PtCl_4^=$ is also an unsolved question.

Even if the actual molecular structures were always known,

the nature of bonding would still be a major problem. At present there is no completely satisfactory way of determining the proportion of covalent to ionic character in a given bond. Nor is there any simple way to show the presence of pi bonding in a molecule. Our understanding of reaction mechanisms would be considerably enhanced if we knew more about the nature of the chemical bond.

Another problem related to the nature of bonding in simple molecules is the elucidation of factors that determine the reactivities of molecules or, in other words, factors that determine bonding in the transition state. To a large extent, bonding in transition states conforms to the same rules as bonding in the ground state, but there is no reason to suppose that activated complexes always follow the simple patterns of normally stable molecules. The high nucleophilic reactivity of peroxide anions is one example of unexplained reactivities of molecules.

Several specific problems yet to be dealt with are detailed below.

10-2 CONSTITUENTS OF A RATE LAW

The rate law for each chemical reaction must be obtained through experiment. Although in many reactions it is possible through analogy to assume the structure of the rate law, theory can predict neither the constitution of the activated complex nor the orders in the law.

These facts are well understood, but one consequence is poorly recognized. The true rate law may differ from the derived law because of the unsuspected presence of some particle. Particularly in radical reactions, the rate law may contain terms for catalysts, sensitizers, or inhibitors. For example, oxidations of organic compounds are very sensitive to the presence of oxygen, even though the observed rate may not be influenced by the further addition of oxygen gas (the result of a plateau of zero-order dependence being initially reached at low pressures of oxygen). A mechanism that overlooks this oxygen dependence is, of course, erroneous.

The chemist interested in understanding a particular reaction mechanism must evaluate the complete rate law. He must discover all factors that affect the rate law and the final rate constant. Some of the unexpected things that can be present in the activated complex are oxygen gas, trace-metal impurities (see

Section 10-3), hydrogen or hydroxide ion, buffer constituents (as in general-acid catalysis), the solvent, and specific ions. It is important that these things be picked up, for they provide significant clues to reaction mechanisms.

10-3 TRACE-METAL CATALYSIS

Early in the studies of peroxide reaction mechanisms, it was recognized that the reactions often had nonreproducible rates. This inconsistency led Bancroft[1] to say that "the chemistry of hydrogen peroxide is a hopeless subject for the phenomenological or Baconian experimenter because misleading experiment is everywhere." Although the author of this text believes the statement to be unduly pessimistic, there is no doubt as to the severity of the problem. Furthermore, reactions of other compounds containing two adjacent electronegative atoms (for example, derivatives of hydroxylamine, hydrazine, and hypohalous acids) are also known occasionally to lack reproducibility in their rates.

It is clear that impurities such as small amounts of transition metal ions are one cause for this rate fluctuation. For example, as little as 10^{-9} M cobaltous ion will increase the rate of decomposition of peroxymonosulfuric acid,[2] and cupric ion is a notorious catalyst for many oxidations by peroxydisulfate ion.

The problem can be divided into two parts. The first part, how to prove that a metal ion is active in a particular reaction, is surprisingly difficult to solve. Whereas a small amount of a metal ion may be very active, addition of more may show little further enhancement. The normal methods of evaluating kinetic orders are ineffective in the study of trace-metal catalysis.

The second part of the problem, how to eliminate the trace-metal catalysis altogether, is also difficult to solve. The addition of a strong complexing agent such as ethylenediaminetetraacetate ion sometimes helps by tying up the metal ion in an inactive form. Unfortunately, most of the strong complexing agents are organic compounds and therefore are sensitive to oxidation by peroxides.

Some reactions that this author believes may be subject to trace-metal catalysis are the decomposition of hydrogen peroxide[3]; the oxidation of hydrogen peroxide by peroxydisulfate ion[4]; and the oxidation of organic and inorganic compounds by peroxydisulfate ion.[5] However, trace-metal catalysis in these and other reactions is difficult to prove or disprove.

10-4 PROBLEMS IN PHOSPHORUS DISPLACEMENTS

The mechanisms for nucleophile reactions with phosphorus substrates are more complicated than they would appear at first sight. In this section, several of the existing problems will be mentioned.

Most phosphorus substrates contain a phosphorus atom surrounded by four groups bonded to the central atom by electrons in sp^3 hybrid orbitals of phosphorus. However, phosphorus also has five unfilled $3d$ orbitals with which an attacking nucleophile may interact. The extent and nature of this interaction is one of the unsolved problems. Since, as seen in Chapter 4, displacement on phosphorus does depend on nucleophile basicity, some interaction between the nucleophile and an open orbital must exist. Looked at more deeply, this problem becomes one of whether a five-coordinate intermediate exists in nucleophilic displacements on phosphorus. On the one hand, interaction of nucleophile and open d orbital may merely be a stabilization of an S_N2 transition state. This would lower the activation energy for the reaction since bond formation would be well developed before bond breaking became significant. On the other hand, the interaction of nucleophile and orbital could be sufficiently strong to give a definite intermediate with a trigonal-bipyramid structure. The reaction of nucleophile N with substrate R_3PX would then be of the type

$$\text{N} + \text{R}_3\text{PX} \underset{k_2}{\overset{k_1}{\rightleftharpoons}} \text{N--P--X} \begin{smallmatrix} \text{R} & & \text{R} \\ & \diagdown \diagup & \\ & | & \\ & \text{R} & \end{smallmatrix}$$

$$\begin{smallmatrix} \text{R} & & \text{R} \\ & \diagdown \diagup & \\ \end{smallmatrix} \text{N--P--X} \begin{smallmatrix} & \\ | \\ \text{R} \end{smallmatrix} \overset{k_3}{\rightarrow} \text{NPR}_3 + \text{X}$$

For such a displacement mechanism (at least for nucleophile exchange where N equals X), microscopic reversibility is assumed to hold. In the intermediate, then, N and X would be symmetrically placed on the axis of the trigonal bipyramid and the three R groups would occupy the equatorial positions.

When $k_3 \gg k_2$, the formation of the intermediate is rate-determining and the rate should depend markedly on the strength of the nucleophile but little on the leaving-group nature. When $k_2 \gg k_3$, the breakdown of the intermediate to give product is rate-determining and leaving-group nature becomes important.

With the more common situation, $k_2 \simeq k_3$, where the energy well caused by the presence of the intermediate is shallow, both bond formation and bond breakage are important as observed.

Attempts to detect the presence of the five-coordinate intermediate have been unsuccessful so far. The observation that no oxygen-isotope exchange occurs during the course of hydrolysis has been considered to rule against an intermediate. Although possibly correct, this consideration is not completely convincing because the incoming nucleophile would be on one axial position of the trigonal bipyramid and the exchanging oxygen would be on the equator. Since these two positions are not identical, exchange might not be detected even if a true intermediate were formed.

In conclusion, it seems certain from compiled nucleophilic orders that some interaction of the open d orbital on phosphorus with the incoming nucleophile does occur. Yet it also seems certain that a five-coordinate intermediate of high stability does not persist in solution. The actual situation, then, lies somewhere between these two extremes; the resolution of exactly where must await further study and evidence.

A second problem is the elucidation of the exact transition-state interactions when a phosphorus compound hydrolyzes in the presence of an acid or a base. Some possible base interactions include:

Specific-base catalysis. This is the usual mechanism found for basic hydrolysis and involves direct interaction of lyate ion with substrate.

$$OH^- + R_3PX \rightarrow HOPR_3 + X^-$$

This mechanism probably holds for many of the observed displacements by hydroxide ion on organophosphorus substrates.

General-base catalysis. In this type of mechanism, any proton acceptor B can interact with a phosphorus substrate to give the activated complex (I). The base B removes a proton from the water

(I)

molecule at the same time that the oxygen attacks the phosphorus. If $B = OH^-$, the rate law is identical to the one for the previous interaction, yet the actual details of the interactions are different.

Nucleophilic catalysis. Although the rate law

$$v = k[\text{B}][\text{R}_3\text{PX}] \tag{10-1}$$

can be ascribed to general-base catalysis of hydrolysis, it also can be ascribed to nucleophilic catalysis. The latter proceeds by the mechanism

$$\text{B} + \text{R}_3\text{PX} \rightarrow \text{BPR}_3 + \text{X}$$

$$\text{BPR}_3 + \text{H}_2\text{O} \rightarrow \text{HOPR}_3 + \text{B} + \text{H}^+$$

where the first step is rate-determining and the second step is a rapid follow-up.

As seen from these three types of interaction, details over and above the rate law must be obtained in order to gain a complete description of a mechanism. Similar ambiguities arise in the acid catalysis of phosphate hydrolyses, where some of the mechanisms are:

Specific-acid catalysis. In this case, the rapid transfer of a proton from hydronium ion to substrate is followed by a rate-determining attack by water. The rate law would be

$$v = k[\text{R}_3\text{PX}][\text{H}^+] \tag{10-2}$$

General-acid catalysis. Here the rate law is

$$v = k[\text{R}_3\text{PX}][\text{HA}] \tag{10-3}$$

where HA is any proton donor. Since the proton must be donated to an atom containing a pair of nonbonded electrons in the transition state, it is probably transferred to the leaving group X. The activated complex would be of the type (II), with a simultaneous water molecule attack on phosphorus and proton transfer.

$$\begin{array}{c} \text{H} \quad \text{R} \quad \text{R} \\ \diagdown \text{O---P---X---HA} \\ \text{H} \diagup \quad | \\ \text{R} \end{array}$$
$$(\text{II})$$

The products of the rate step are, therefore, H_2OPR_3^+, HX, and A^-. Essentially, the general acid is aiding the cleavage of the P—X bond by helping to withdraw electrons from this bond.

Specific-acid and general-base catalysis. The rate law [Eq. (10-3)] characteristic of general-acid catalysis is also observed if an equilibrium proton transfer to substrate is followed by a rate-determining general-base (or nucleophilic) catalysis.

Problems like the disposition of protons or buffer constituents in the activated complex are, in principle, solvable using kinetic isotope effects. To date, however, attempts to solve these mechanistic problems have been limited in scope, and much remains to be learned about displacements on tetrahedral phosphorus.

10-5   REPLACEMENTS IN COORDINATION COMPOUNDS

In Chapter 6, some of the problems faced by the chemist studying ligand replacement in transition metal complexes were mentioned. Here it is necessary only to summarize and to illustrate the basic problem, that of discovering the extent of bond formation to incoming ligands.

To date, kinetic studies have helped little because the reactions fall into two main classes: (a) aquations that, as a result of unit solvent activity, appear first-order even when bimolecular; and (b) second-order reactions whose mechanisms are still unresolved because their kinetics are consistent with either an S_N2 displacement or a fast reversible loss of a bound ligand prior to a rate-determining step involving intermediate and incoming ligand. Another factor that influences the kinetics is ion pairing between a positively charged complex and an anionic ligand.

This problem of discovering whether reactions of complexes are S_N1 or S_N2 has recently been reinvestigated by two groups of chemists. Haim and Wilmarth[6] have investigated replacements in negatively charged complexes of the type $[Co(CN)_5L]^=$, where the problem of ion pairing to anionic ligands is eliminated. The pseudo-first-order rate constants for substitution of H_2O by N_3^- or SCN^- in $[Co(CN)_5OH_2]^=$ fit the equation

$$k = \frac{k_1[X^-]}{[X^-] + k_2/k_3} \tag{10-4}$$

where X^- is a nucleophilic anion. The nature of this rate law along with other evidence strongly supports a limiting S_N1 mechanism involving the pentacoordinated intermediate $[Co(CN)_5]^=$.

By way of contrast, a replacement in an octahedral complex that almost certainly proceeds by an S_N2 mechanism has recently come to light. The hydrolysis of tris(acetylacetonato)silicon(IV) cation is subject to catalysis by a variety of nucleophilic ligands, and the relative rates conform to a reasonable order of nucleophilic strengths.[7] It should be noticed that the central atom, silicon, has empty $3d$ orbitals available for bonding to the incoming nucleophile.

The two cases just mentioned are, without doubt, extremes, one involving a typical $S_N 1$ mechanism and the other involving a straightforward displacement step. The majority of cases lie somewhere between, and it is a recurrent problem to evaluate the extent of bond formation in such cases.†

10-6 NITROSATION REACTIONS

The general reaction by which an NO^+ moiety from nitrite ion attaches itself to an electron donor Y to form the particle YNO^+ is called nitrosation. The kinetics of a large number of such nitrosation reactions have been studied; yet the actual intermediate that transfers the NO^+ group to Y in the rate step is still uncertain. Some rate laws and possible conclusions will be mentioned here.

The exchange of oxygen atoms between nitrite ion and solvent water was discussed in Chapter 8. The characteristic law for oxyanion exchanges

$$v = k[NO_2^-][H^+]^2 \tag{10-5}$$

was found by Anbar and Taube.[8] Under somewhat different conditions,[9] the law

$$v = k[HNO_2]^2 \tag{10-6}$$

was obtained, and the rate constant is similar to the one estimated from the rate of diazotization of aniline. In the presence of acetate ion, exchange can proceed by a mechanism involving acetate ion as catalyst.[10]

The reaction of nitrite ion with amines was recently reviewed by Ridd,[11] who discussed the processes of nitrosation, diazotization, and deamination from a common point of view—the mechanism

---

† Some of the complexities and difficulties in dealing with reactions of coordination compounds are exemplified in the careful studies of C. H. Langford and P. Langford, *Inorg. Chem.*, **2**, 300 (1963) and of D. J. MacDonald and C. S. Garner, *Inorg. Chem.*, **1**, 20 (1962).

of the nitrosation step. A variety of rate laws are observed, among them the following:

$$v = k[RNH_2][HNO_2][H^+] \tag{10-7}$$

$$v = k[RNH_2][HNO_2][H^+][X^-] \tag{10-8}$$

$$v = k[RNH_2][HNO_2]^2 \tag{10-9}$$

$$v = k[HNO_2][H^+][X^-] \tag{10-10}$$

$$v = k[HNO_2]^2 \tag{10-11}$$

where X^- is any one of a series of such anions as acetate, phthalate, chloride, bromide, thiocyanate, and iodide ion. The rate laws that are second-order in nitrous acid concentration [Eqs. (10-9) and (10-11)] are thought to indicate the intermediacy of N_2O_3 formed by the reaction

$$2HNO_2 \rightleftharpoons N_2O_3 + H_2O$$

in the nitrosation media. The rate laws that are first-order in X^- [Eqs. (10-8) and (10-10)] are thought to indicate the intermediacy of an XNO particle.† Rate laws that are first-order in amine substrate concentration [Eqs. (10-7), (10-8), and (10-9)] represent attack by an electrophilic intermediate ($H_2NO_2^+$, N_2O_3, or XNO) at the spare pair of electrons on the amine nitrogen. On the other hand, rate laws that are zero-order in amine concentration [Eqs. (10-10) and (10-11)] represent rate-determining formation of the intermediate. Although he considers the nitrosonium ion NO^+ to be a nitrosating intermediate in strong acid, Ridd[11] believes it to be inactive in weakly acid solutions ($pH \geq 4$) where many nitrosations occur at convenient rates.

In contrast, the results of Anbar and Taube[8] suggest the formation of NO^+ in dilute aqueous acid. With hydrogen peroxide as the substrate, the rate law at low peroxide concentrations is

$$v = k[H_2O_2][NO_2^-][H^+]^2 \tag{10-12}$$

† However, the reader should in general be wary of this type of conclusion. Nitrosation reactions proceed by mechanisms involving several steps, and slight variation in conditions or in substrate can influence the rates of these steps so that one or another may be rate-determining. Indeed in one case, the nitrosodeprotonation reaction, the rate step was found to be proton loss from the aromatic ring after addition of the nitroso group; therefore in this case, the rate law gives no information as to the nature of the nitrosating species [K. M. Ibne-Rasa, *J. Am. Chem. Soc.*, **84**, 4962 (1962)].

and at high peroxide concentrations is

$$v = k[NO_2^-][H^+]^2 \tag{10-13}$$

Since the presence of hydrogen peroxide retards the exchange of oxygen between nitrite ion and water, hydrogen peroxide and water must be competing for the same electrophilic intermediate. Further, the rate constant for the law at high peroxide concentrations is the same as that found for oxygen exchange under similar conditions.

Undoubtedly there are a variety of pathways by which nitrosation can occur. At least in part, differing reports from various investigators result from the lack of identical reaction conditions, for a slight variation of conditions is sufficient to cause one mechanism to predominate over another. The voluminous literature on the nitrosation mechanism is highly argumentative, but the situation is still not resolved. A particularly important point at issue is whether the nitrosonium ion is indeed an electrophilic intermediate in the nitrosation of electron-pair donors in dilute aqueous acid.

10-7　SOME REACTIONS TO STUDY

In the course of his research, every chemist comes across reactions whose mechanisms are worth studying. The following are a few that this writer has encountered but has had no chance to study in detail.

The *ionization of boric acid*

$$B(OH)_3 + 2H_2O \rightleftharpoons B(OH)_4^- + H_3O^+$$

is interesting because the boron changes coordination number from three to four. The reaction is known to be fast, for no delay is observed in the course of titrations. Although the kinetics of this unusual process cannot be studied by normal techniques, the newer techniques for fast reactions[12] should prove helpful. For example, nuclear magnetic resonance of the boron-11 isotope or the oxygen-17 isotope might be appropriate.

In the course of investigating the kinetics of *tellurate ion complexing by polyols*,[13] it became apparent that the mechanism involves a tellurium intermediate, possibly $TeO_4^=$. This intermediate was postulated to be formed prior to the rate-determining step. Such a postulation requires that the rate of oxygen exchange between tellurate ion $H_5TeO_6^-$ and water be more rapid than the complexing reaction itself. Data to test this prediction are not available.

The *oxidation of thiosulfate ion by nitrite ion in weakly acid solution* was once briefly investigated by the kinetic technique.[14] The rate law reported was

$$v = k[NO_2^-][H^+]^2[S_2O_3^=]^{0.2} \tag{10-14}$$

It is interesting because the order in reducing agent is nearly zero. This suggests that the mechanism is related to those for the nitrosation of aniline and for the exchange of nitrite oxygen with water. A careful study of the stoichiometry and mechanism is still needed.

References

1. W. D. Bancroft and N. F. Murphy, *J. Phys. Chem.*, **39**, 377 (1935).
2. D. L. Ball and J. O. Edwards, *J. Phys. Chem.*, **62**, 343 (1958).
3. F. R. Duke and T. W. Haas, *J. Phys. Chem.*, **65**, 304 (1961); E. Koubek, M. L. Haggett, C. J. Battaglia, K. M. Ibne-Rasa, H. Y. Pyun, and J. O. Edwards, *J. Am. Chem. Soc.*, **85**, 2263 (1963).
4. M. Tsao and W. K. Wilmarth, *Discussions Faraday Soc.*, **29**, 137 (1960).
5. See discussion by W. K. Wilmarth and A. Haim in J. O. Edwards (ed.), *Peroxide Reaction Mechanisms*, Wiley-Interscience, New York, 1962, p. 175–225.
6. A. Haim and W. K. Wilmarth, *Inorg. Chem.*, **1**, 573, 583 (1962).
7. R. G. Pearson, D. N. Edgington, and F. Basolo, *J. Am. Chem. Soc.*, **84**, 3233 (1962).
8. M. Anbar and H. Taube, *J. Am. Chem. Soc.*, **76**, 6243 (1954).
9. C. A. Bunton, D. R. Llewellyn, and G. Stedman, *J. Chem. Soc.*, **1959**, 568.
10. C. A. Bunton and H. Masui, *J. Chem. Soc.*, **1960**, 304.
11. J. H. Ridd, *Quart. Rev. (London)*, **15**, 418 (1961).
12. A. A. Frost and R. G. Pearson, *Kinetics and Mechanism*, 2d ed., Wiley, New York, 1961, Chap. 11.
13. H. R. Ellison, J. O. Edwards, and L. Nyberg, *J. Am. Chem. Soc.*, **84**, 1824 (1962).
14. J. O. Edwards, *Science*, **113**, 392 (1951).

Index

Index